Plant Peroxisomes

AMERICAN SOCIETY OF PLANT PHYSIOLOGISTS MONOGRAPH SERIES

Anthony H. C. Huang, Richard N. Trelease, and
Thomas S. Moore, Jr. *PLANT PEROXISOMES, 1983.*

Plant Peroxisomes

Anthony H. C. Huang
Department of Biology
University of South Carolina
Columbia, South Carolina

Richard N. Trelease
Department of Botany and Microbiology
Arizona State University
Tempe, Arizona

Thomas S. Moore, Jr.
Department of Botany
Louisiana State University
Baton Rouge, Louisiana

1983

ACADEMIC PRESS
A Subsidiary of Harcourt Brace Jovanovich, Publishers
New York London
Paris San Diego San Francisco São Paulo Sydney Tokyo Toronto

ACADEMIC PRESS, INC.
111 Fifth Avenue, New York, New York 10003

United Kingdom Edition published by
ACADEMIC PRESS, INC. (LONDON) LTD.
24/28 Oval Road, London NW1 7DX

Library of Congress Cataloging in Publication Data

Huang, Anthony, H. C.
Plant Peroxisomes

(American Society of Plant Physiologists monograph series)
Includes index
1. Plant cell microbodies. I. Trelease, Richard N.
II. Moore, Thomas S., Jr. III. Title. IV. Series
QK725.H77 1983 581.87'34 82-22777
ISBN 0-12-358260-1

PRINTED IN THE UNITED STATES OF AMERICA

83 84 85 86 9 8 7 6 5 4 3 2 1

Contents

Foreword

This is the first book in a new series sponsored by the Monograph Committee of the American Society of Plant Physiologists, an ongoing committee created to encourage qualified authors to write definitive books on important subjects in order to further the science and profession of plant physiology. *Plant Peroxisomes* fulfills the goals of the Committee, sets a high standard of scholarship for the series, and comes at an opportune time. The presence of these organelles and some of their functions in plant tissues were firmly established about 15 years ago, yet there are still important questions to be understood about the development and regulation of enzymes in peroxisomes involved in important plant processes, including photorespiration.

Huang, Trelease, and Moore are scientists who have made original research contributions to the subject. They have succeeded in producing a scholarly and readable volume in which the chapters and the style are so well integrated that one is surprised to realize that it contains contributions from three individual authors.

The book begins with a historical summary that demonstrates how advances in electron microscopy and cell fractionation studies have led to the detection of these fragile organelles (first in mouse kidney cells); the summary is followed by a description of the types of plant peroxisomes and some of their metabolic functions. There is a thorough discussion of cytochemical procedures for establishing the localization of specific enzymes and for an analysis of the distribution of peroxisomes, including those in nonflowering plants, algae, and fungi. A useful section describes methods of isolating peroxisomes from various tissues and the physical and chemical properties of the isolated organelles.

The largest part of the monograph is devoted to a discussion of the metabolic functions of peroxisomes, their enzymatic activities, and the development of their form and biochemistry. It is in this area that the exciting research discoveries of the future will likely occur, and the authors provide a sound foundation for further study. Among the important functions of peroxisomes is their role in gluconeogenesis from reserve lipids in germinating oil-rich seeds and in photorespiration in leaves of C_3 plant species. Several key enzymes of the glycolate pathway of photorespiration are found in leaf peroxisomes, and the present knowledge about their activities is presented clearly and accurately. Because peroxisomes appear to contain no DNA, the synthesis of enzymes and organelle development is presumably controlled by the nucleus, but little information now exists about the genetic control of the organelle or its enzymes. A helpful feature of the book is the summary at the end of each chapter, which may encourage the hasty reader to delve further into the fascinating details that precede it. The authors have also provided a concluding perspective describing what they consider to be the major gaps in knowledge about peroxisomes and their physiology.

This book should be of value to professional scientists concerned with plant metabolism and development, as well as graduate students.

Israel Zelitch

Preface

The Monograph Committee of the American Society of Plant Physiologists in 1978 recommended that the three of us, with varied backgrounds and orientations in peroxisome research, write a monograph on plant microbodies (peroxisomes). We felt we could satisfactorily cover the field in a monograph format, but then considered whether we would contribute new information or provide better insight than the several review articles and a monograph (albeit in German) that had just been published. There were several important aspects of plant peroxisomes that had not been covered or thoroughly integrated. For example, a comprehensive discussion on the historical perspective of peroxisomal studies had not been presented. In addition, there was a need for describing and explaining the various types of cytochemical tests used successfully to help define the function and distribution of these organelles in plant cells. Also, a review on the biochemical or metabolic aspects of fungal and algal peroxisomes had not been written. Most of the descriptions on glyoxysomes in the available reviews centered on castor bean, and diversions from this model system in other organisms (other oilseeds, fungi, and algae) have not been described sufficiently. Finally, the most current information on the biogenesis of plant peroxisomes and its relevance to the proposed model for peroxisome biogenesis in germinated seeds needed to be assembled and assessed. Thus, it became apparent that a monograph containing the above information, as well as comprehensive descriptions and evaluations of data on other aspects, written by joint authorship would be beneficial to plant physiologists and other biologists in related fields.

During the 3-year (1980 to late-1982) period of writing and reviewing, there were significant advances in several areas of peroxisome research. Two of these areas are (1) the metabolic involvement of peroxisomes in ureide metabolism (in nodulated roots) and (2) the mode of peroxisome (especially glyoxysome) biosynthesis and differentiation. A timely international symposium on animal and plant peroxisomes emphasizing fatty acid β-oxidation in animal peroxisomes and the synthesis, intracellular transport, and packaging of peroxisome proteins in both animal and plant peroxisomes was held in September 1981. The essence of these findings is included. We hope that the comprehensive nature of our coverage will illustrate the significance of plant peroxisomes and their roles in the physiology and metabolism of plants.

We would like to make a special note on the title. The committee asked that we write a monograph entitled "Plant Microbodies." After completion of our first draft, we concluded that this was inappropriate. Our first thought for a new one was "Plant Microbodies (Glyoxysomes and Peroxisomes)," but this would have helped to perpetuate the microbody term and was redundant. Then we considered "Plant Peroxisomes and Glyoxysomes," but this title implies that glyoxysomes are not peroxisomes. The title used by Gerhardt for his 1978 monograph—"Microbodies/Peroxisomes in Plant Cells"—at least implies that microbodies are peroxisomes, but reinforces calling these organelles microbodies. Finally, we reduced the title to "Plant Peroxisomes," one that we felt encompasses *all types* of microbodies, including glyoxysomes. Our main reluctance to this choice was the concern that readers would think, because of previous experiences with the terminology, the monograph dealt only with organelles in green tissues and not glyoxysomes and other microbodies (peroxisomes). We hope readers will understand and agree that the title we ultimately selected is most appropriate.

The Monograph Committee specifically indicated that the monograph should not be a detailed review of the literature. Thus, only selected references are included in this volume. Specifically, we have included in Chapter 2 several tables that do not cite specific references for each of the numerous listings. Citations to articles published prior to 1978 can be found in similar tables in B. Gerhardt's monograph (Microbodies/Peroxisomen pflanzlicher Zellen, *Cell Biology Monographs,* Vol. 5). Whenever possible, high quality electron micrographs, not published previously, were used to illustrate peroxisomes in various plant cells. This was done to give the reader a refreshing view of peroxisome ultrastructure, because original micrographs from published papers have been used repetitiously in many review articles.

We wish to extend our sincerest thanks and gratitude to those colleagues, too numerous to list here, who have reviewed parts of our manuscripts and contributed figures and tables of their published and unpublished work. However, a special acknowledgment is due Dr. P. Gruber for his critical review and contributions to Chapters 1 and 2. We also thank the National Science Foundation and the U.S.D.A. Competitive Research Grant Program for their generous support of our research on peroxisomes in recent years. The secretarial assistance of Ms. Beate Christy is greatly appreciated.

Anthony H. C. Huang
Richard N. Trelease
Thomas S. Moore, Jr.

1

Introduction

I. CURRENT TERMINOLOGY

Microbodies are a class of ultrastructurally defined organelles with diameters ranging from 0.2 to 1.7 μm. They contain a coarsely granular or fibrillar matrix (occasionally with amorphous or paracrystalline inclusions) bounded by a single membrane. The term *microbody* is a general name, not implying any particular function. Microbodies in animal cells known to contain at least catalase and a H_2O_2-producing oxidase are termed *peroxisomes* (de Duve, 1965). This term was conceived to emphasize the role of microbodies in hydrogen peroxide metabolism. It does *not* imply that these organelles contain the enzyme peroxidase; in fact, no true peroxidase has ever been found in peroxisomes. In plant cells, electron microscopic observations have revealed a widespread occurrence of organelles morphologically defined as microbodies. Sufficient cytochemical and biochemical evidence for their possession of catalase and at least one H_2O_2-producing oxidase has been accumulated to allow categorizing plant microbodies as peroxisomes.

One main class of plant peroxisomes, commonly found in oil-rich tissues of seeds, contains fatty acid β-oxidation and glyoxylate cycle enzymes in addition to the characteristic peroxisomal enzymes. These peroxisomes are called *glyoxysomes*. Another class, commonly found in green leaves and cotyledons, contains enzymes that oxidize and process glycolate as part of the photorespiration process. These have been called *leaf peroxisomes* for brevity, although a term such as "leaf-type" peroxisomes would more aptly describe all those peroxisomes of similar characteristics occurring in leaves, green cotyledons, and special leaf tissues in some C_4 species. In some higher-plant tissues and in many algal and fungal species, the peroxisomes are known to be involved in metabolism of compounds such as urate, methanol, amines, and oxalate. Although these peroxisomes are highly specialized to perform unique physiological functions, they have not been given special names; they are simply called peroxisomes based on their possession of the characteristic enzymes. Catalase-containing peroxisomes in tissues or cells not involved in any of the above physiological processes are refered to as *unspecialized peroxisomes*. They are denoted as "unspecialized" only because their role in plant cell metabolism has not yet been elucidated.

Glyoxysomes, leaf peroxisomes, peroxisomes for other special metabolism, and unspecialized peroxisomes will be the terms used throughout this monograph in the context described above. The term microbody will be used in a general sense to refer to any or all of the above peroxisomes, or it will be used to describe such organelles observed with the electron microscope which have not been tested for their enzyme content.

II. HISTORICAL PERSPECTIVE

A. Ultrastructural Discoveries

Peroxisomes were not originally discovered in cell fractionation experiments, but were first described as microbodies in electron microscopic studies of thin cell sections. Rhodin (1954) used the term "microbody" to describe the appearance of a single membrane-bounded organelle present in the convoluted tubule cells of mouse kidney. Two years later, a similar morphological body, but with a regular crystalloid structure within its matrix, was identified in rat liver cells (Rouiller and Bernhard, 1956). Results from cell fractionation experiments showed that the microbodies in liver and kidney cells, separated from organelles containing hydrolases (lysosomes), contained urate oxidase, catalase, α-hydroxyacid oxidase, and D-amino acid oxidase activity. To emphasize the biological significance of the oxidative enzyme activity and the generation of hydrogen peroxide, de Duve (1965) coined the term "peroxisome" to define this animal cell organelle more specifically. The distribution of peroxisomes in numerous other animal tissues was made known following the discovery of much smaller (0.22–0.86 μm diameter) catalase-containing microbodies in neural cells which Novikoff and Novikoff (1972) appropriately named "microperoxisomes." Peroxisomes are now considered to have a ubiquitous distribution among the myriad of mammalian cells, and most likely are common among other vertebrate and invertebrate cells.

The early discovery and electron microscopic description of microbodies in plant cells were not as straightforward. Similar organelles in plant cells were not equated with animal microbodies until some 12 years after Rhodin's original description. This impasse can be attributed to the initial difficulties in preserving plant material for electron microscopic observation and the incorrect assignment of names to the organelles seen with the electron microscope. A concerted effort was made to preserve terminology originating from classical light microscopic literature on plant cells. This led to confusing and overlapping naming of organelles, some of which still persists today.

In the 1950s, buffered osmium tetroxide (OsO_4) served as a good fixative for preserving ultrastructural detail of animal cells, but generally was inadequate for plant cells. The plant cytologists had to rely on potassium permanganate ($KMnO_4$) as a reasonable preservative. Although much information was gained on plant cell ultrastructure, the images produced were quite dissimilar from the OsO_4-preserved images of animal cells. $KMnO_4$ oxidizes most cellular components, leaving a "membrane outline" of the various organelles. The trend for plant cytologists to ultrastructurally define and account for all light microscopic structures led to attributing many names and functions to permanganate-fixed images of microbodies. The irony is that microbodies (peroxisomes) as known today probably were never described with the optical microscope.

A pertinent example of this carry-over was assignment of the name "spherosome" to ultrastructural images. Small refractile bodies that shined in dark-field optical microscopy were the subject of numerous studies on Brownian movement and protoplasmic streaming in the early 1900s. Nearly 100 years ago Hanstein termed these structures "microsomes," but the name was relinquished in 1953 following Albert Claude's thorough description of microsomes as endoplasmic reticulum fragments from fibroblast cells (he was unaware of the earlier botanical use of the term). Spherosomes presumably were lipoidal in nature (since they readily stained with lipid dyes) and were thought to contain protein owing to the refractile qualities and positive reactions with certain redox dyes. Figure 1.1 is a representative micrograph of plant cells fixed with $KMnO_4$. Plastids, mitochondria, nuclei, endoplasmic reticulum, and Golgi described from optical studies were clearly resolved and correctly identified in micrographs. Two structures commonly seen in various cell types (labeled correctly as "mb" and "ob" in Fig. 1.1) were not readily identifiable by electron microscopists in the early 1960s. Both appeared to have a single limiting membrane. Their difference in electron density indicated they were different organelles with different functions. However, in this infant period of electron microscopy, both structures were either ignored or, when labeled, marked as unidentified bodies. A notable exception was that when the organelles having the appearance of "mb" (microbodies) in Fig. 1.1 were viewed in the cell plate region of dividing cells, they received considerable attention under the name "phragmosomes" (Porter and Caulfield, 1958; Manton, 1961). Today, the phragmosomes are recognized as microbodies (peroxisomes) (Hanzely and Vigil, 1975).

Being knowledgeable of the early optical microscopic literature on spherosomes, Frey-Wyssling and his colleagues attempted to resolve the controversy of whether spherosomes were organelles with a special structure and function or simply inert bodies of "ergastic" material. Toward this end, they did an electron microscopic study of onion epidermal and corn coleorhiza cells which were used previously to study spherosomes by optical microscopy. The two organelles shown in Fig. 1.1 as microbodies and oil bodies were identified by Frey-Wyssling *et al.* (1963) as spherosomes and oil bodies, respectively. Their ultrastructural observations and consideration of information derived from optical microscope studies led them to propose an ontogenetic relationship between "spherosomes" (actually microbodies) and oil bodies. A summary of their proposed scheme is shown in Fig. 1.2. Spherosomes appeared to originate from ER, and then to differentiate into oil bodies (ob) by losing their electron opacity because of an accumulation of electron-transparent lipid material (triacylglycerol). The matrix material within spherosomes was thought to be proteinaceous, partially composed of lipid-synthesizing enzymes. Numerous histochemical studies showing particular hydrolase activity in plant cells, and de Duve's earlier demonstration of single-membrane-bound lysosomes in animal cells, led many cytologists to conclude that spherosomes also functioned as plant lysosomes. Thus, 9 years after Rhodin's description of animal microbodies, plant cytologists had identified plant microbodies as hydrolase-containing spherosomes ("S" in Fig. 1.2) destined to differentiate into oil bodies.

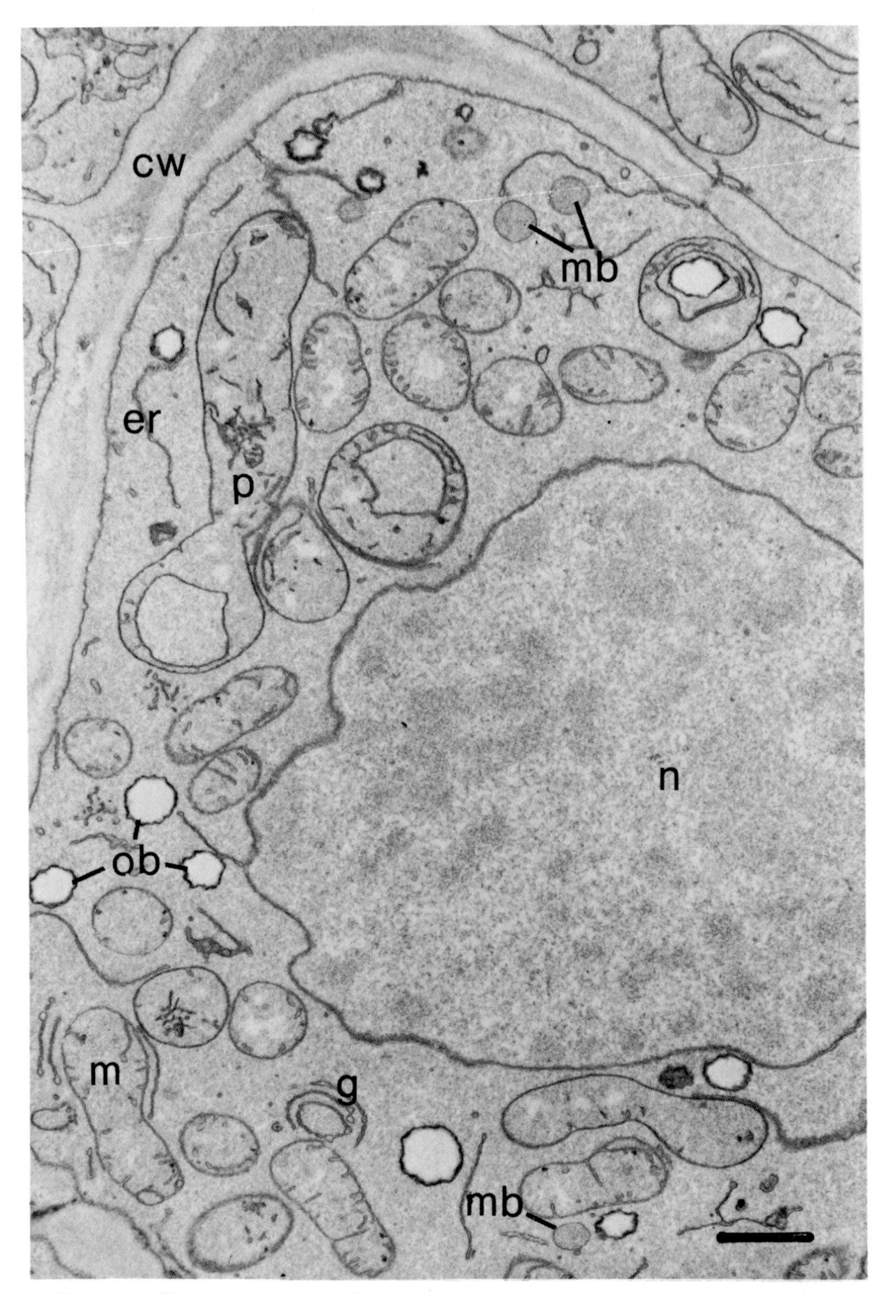

Fig. 1.1. Illustration of part of a meristematic cell in the shoot apex of an 8-day-old corn seedling fixed with 2% $KMnO_4$ for 2 hr. Essentially all the typical cell components that appear in plant cells fixed with $KMnO_4$ are shown. m, mitochondrion; p, plastid; n, nucleus; g, Golgi body; er, endoplasmic reticulum; cw, cell wall; mb, microbodies (peroxisomes); ob, oil bodies. Microbodies and oil bodies are labeled according to present-day knowledge. Bar = 1 μm. From R. N. Trelease (unpublished).

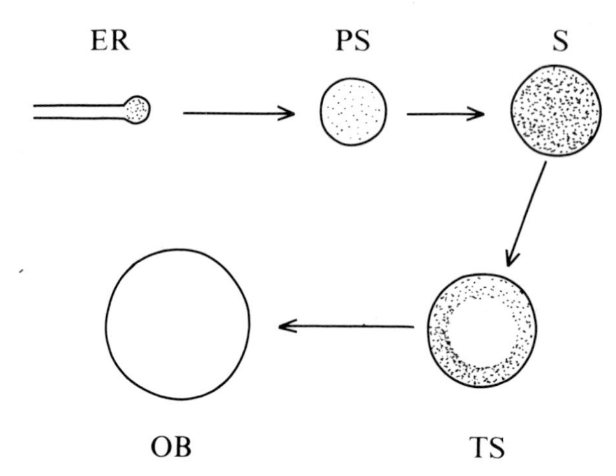

Fig. 1.2. Proposed scheme for the development of spherosomes and oil bodies. Small membrane vesicles pinch off from the endoplasmic reticulum (ER) to form a prospherosome (PS), which differentiates further to become a spherosome (S). Spherosomes go through a transition stage (TS) to finally become oil bodies (OB) with an electron-transparent matrix. Adapted from Frey-Wyssling *et al.* (1963).

A decided impact on plant cell biology was the introduction of glutaraldehyde in 1963 as a primary fixative for electron microscopy. The image of organelles in plant cells became comparable with glutaraldehyde-fixed organelles in animal cells. In a comprehensive study of various cell types from numerous plant species, Mollenhauer *et al.* (1966) made the observation (in mostly $KMnO_4$- and some glutaraldehyde-fixed tissues) that single-membrane-bound organelles characterized by a close association with ER (for examples, see Fig. 1.3) bore a marked structural similarity with the organelles originally described in 1954 as microbodies in animal cells (Fig. 1.4). They proposed that these plant cell "cytosomes" resembling animal microbodies should collectively be called "plant microbodies." It should be noted that at the annual meeting for American Society for Cell Biology in 1965 (1 year before the appearance of Mollenhauer's article), de Duve had christened the microbodies in rat liver and kidney cells as peroxisomes.

Fig. 1.3. (Top) Microbodies (unspecialized peroxisomes) in cells of the bean root tip. Cisterna of RER with ribosomes on the surface away from the upper peroxisome is evident. Bar = 0.25 μm. From R. N. Trelease (unpublished). (Middle) Microbodies (unspecialized peroxisomes) in a parenchyma cell of wheat coleoptile. RER runs along the upper peroxisome containing several noncrystalline inclusions. Bar = 0.25 μm. From R. N. Trelease (unpublished). (Bottom) Elongate microbody (leaf peroxisome) in a palisade cell of a 5-day-old bean leaf. RER is observed on both sides of this profile view. m, mitochondria; p, plastid. Bar = 0.5 μm. From P. J. Gruber (unpublished).

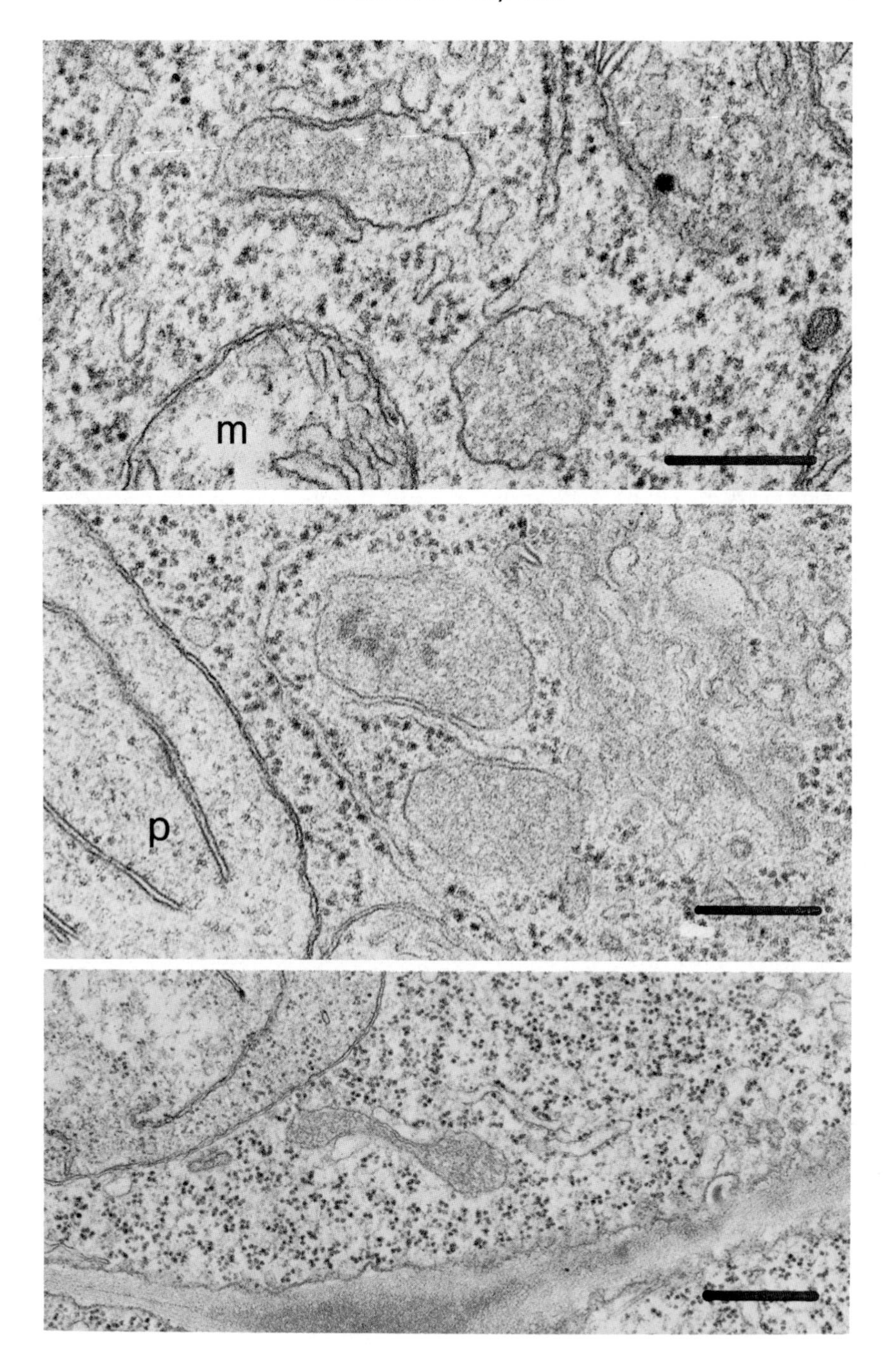
m
p

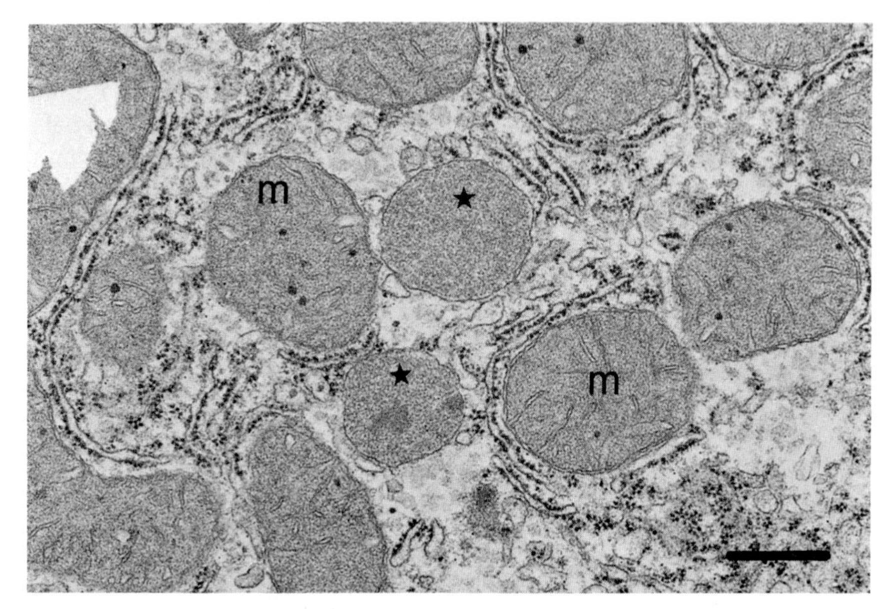

Fig. 1.4. Microbodies (peroxisomes) (stars) and mitochondria (m) in a parenchyma liver cell of a rat. A single membrane limits the peroxisomes; the lower peroxisome contains amorphous dense inclusions. Note close associations of peroxisomes with ER and similar morphology to plant peroxisomes shown in Fig. 1.3. Bar = 0.5 μm. From J. R. Swafford (unpublished).

Mollenhauer *et al.* (1966) also observed dense, noncrystalline inclusions similar to those described in animal microbodies, but they did not observe any crystalline inclusions in plant microbodies. Thus, the crystal-containing bodies (CCBs) first described in several plant tissues by Cronshaw (1964) and in oat coleoptile by Thorton and Thimann (1964) were still not acknowledged as microbodies. Mollenhauer *et al.* believed that the phragmosomes occurring in the cell plate region of dividing cells were microbodies, and they distinguished spherosomes from microbodies because spherosomes "seem to represent lipid bodies and are not known to be endoplasmic reticulum associated." They did not consider microbodies to be the site of hydrolase activity, since lysosomes were distinct from microbodies in animal cells. However, an alternative site of lysosomal activity was not proposed, creating further confusion as to which plant cell organelle(s) was responsible for the lytic activities detected in cell-free extracts and localized in sections with the optical microscope.

Two years later, Newcomb and co-workers published a comprehensive report aimed at clarifying the fine structural properties of plant microbodies in tissues fixed with glutaraldehyde and OsO_4 (Frederick *et al.*, 1968). They concurred that microbody was a good term, since it did not imply any particular function and essentially no data existed on the enzyme content of these plant organelles. In contrast to Mollenhauer's findings, regular crystalline arrays often were observed within microbodies (Fig. 1.5). Crystal containing bodies were considered to merely represent a specialized type of microbodies. The only organelle that could be confused with microbody was a homogeneous, low-contrast spherical body not clearly bounded by a unit membrane or associated with endoplasmic reticulum. This body was considered to be a spherosome (Fig. 1.6), because its morphological features were consistent with the lipid bodies, called spherosomes, isolated and characterized from oilseeds by Jacks *et al.* (1967). Thus, the more extensive ultrastructural studies of plant cells led to substantial modification of Frey-Wyssling's ontogenic scheme (Frey-Wyssling *et al.*, 1963) (Fig. 1.3). Oil bodies were called spherosomes (as they often are today) and the hypothetical progenitors of oil bodies (called spherosomes by Frey-Wyssling) were cor-

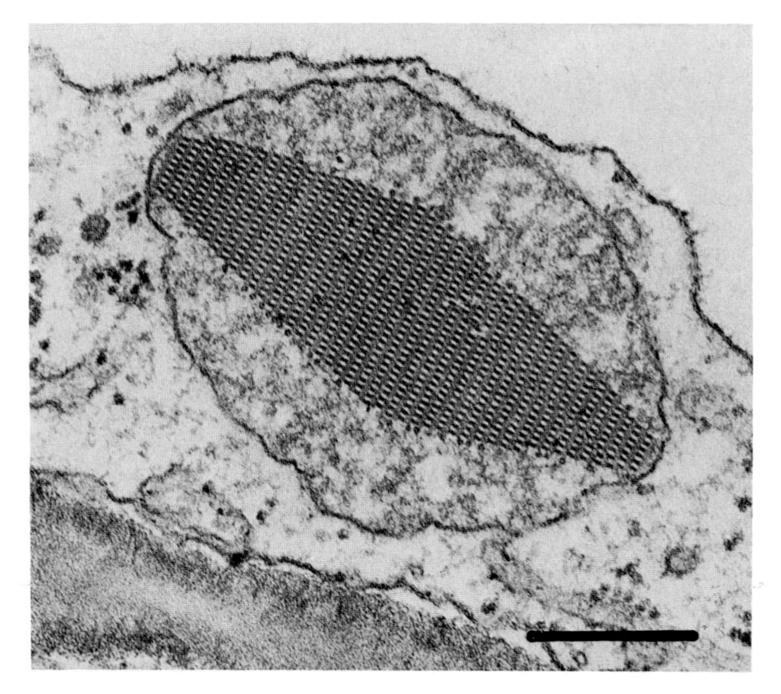

Fig. 1.5. Microbody (unspecialized peroxisome) (CCB) in a parenchyma cell of oat coleoptile. A large single crystal occupies much of the peroxisome matrix. Bar = 0.25 μm. From Frederick *et al.* (1968); courtesy of E. L. Vigil.

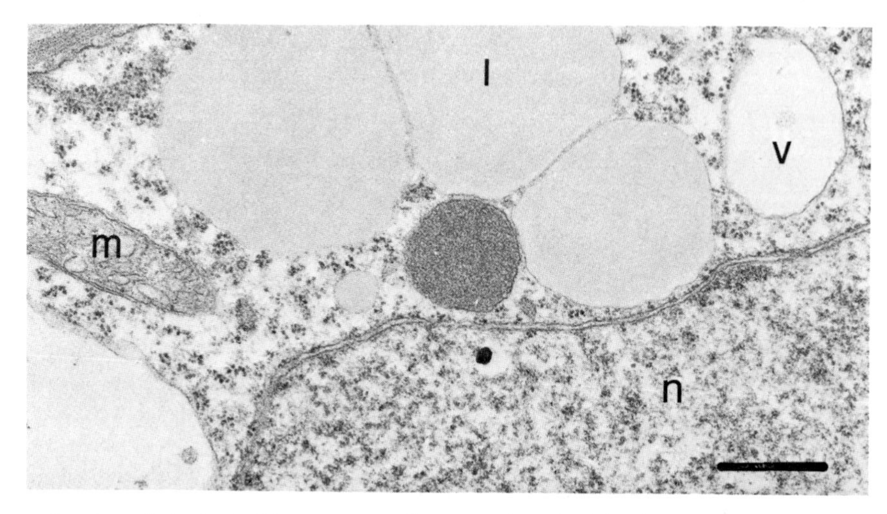

Fig. 1.6. Microbody (glyoxysome) adjacent to three lipid bodies (spherosomes) in an epidermal cell of a cucumber cotyledon (3-day-old seedling). The lipid body was the only organelle observed in glutaraldehyde-fixed cells which could be confused with microbodies. Two membranes are clearly visible around mitochondrion (m), the nucleus (n), and one membrane bounds the glyoxysome and vacuole (v). Lipid bodies have a faint, but distinct boundary layer which has been characterized as a "half unit-membrane." Bar = 0.5 μm. From R. N. Trelease (unpublished).

rectly identified as microbodies, ontogenetically unrelated to oil bodies. Still, the distinction between microbodies and oil-storing spherosomes left unanswered the site of hydrolase activity. In view of the accumulating data on plant hydrolase activity, Frederick *et al.* (1968) stated "the conclusion is almost inescapable that the particles now bearing other names in some of the enzyme localization studies will prove to be identical with microbodies." Hence, in the late 1960s the possibility remained that plant microbodies were lysosomal in nature or, alternatively, changed their enzyme complement such that both hydrolases and oxidases occurred in the same organelles at one stage or another. It seemed entirely possible that a sharp distinction between microbodies (peroxisomes) and lysosomes did not exist in plants as in animals. Although the ultrastructure of microbodies *in situ* had been defined, careful monitoring of biochemical fractions with the electron microscope clearly was needed to further clarify plant microbody function.

B. Biochemical Elucidation of Microbody Function

The first functional role attributed to microbodies in plants came from the work of Beevers and associates. In 1957, Kornberg and Beevers discovered that glyoxylate cycle enzymes were pivotal in the conversion of storage lipid to sucrose during germination of oilseeds (such as castor bean). Numerous studies during the ensuing 10 years suggested that the glyoxylate cycle was operative within the mitochondria (the so-called "particulate" fraction). This proposed localization of enzymes presented a problem as to how carbon flow was regulated between the glyoxylate and Krebs cycles within the same organelle. In 1967, Breidenbach and Beevers centrifuged a particulate fraction on a sucrose gradient and found novel particles with a complete complement of glyoxylate cycle enzymes clearly separated at a higher equilibrium density from the mitochondria. This distinct population of organelles was referred to as "glyoxysomes" to emphasize their role in glyoxylate cycle metabolism. Subsequent electron microscopic examination of the glyoxysome fraction revealed that the "novel" particles were morphologically similar to the microbodies defined by Mollenhauer *et al.* (1966). Moreover, the glyoxysomes were found to contain catalase and glycolate oxidase activity, characteristic of peroxisomes isolated from mammalian and certain protozoan cells. Although they are a type of peroxisome, the name glyoxysome has been retained to recognize their distinct function in plant metabolism. Glyoxysomes were observed earlier with the electron microscope in yucca perisperm (an oil-storage tissue); the organelles were not recognized as microbodies, but were referred to as "unidentified cytoplasmic organelles" (UCOs) (Horner and Arnott, 1966).

Directly following Beevers' elucidation of the function of microbodies in oilseeds, the role of microbodies in leaves was clarified. As part of their ongoing research on glycolate metabolism in photosynthetic tissues, Tolbert's group (Tolbert *et al.*, 1968) discovered that microbodies isolated on sucrose gradients from leaf extracts contained the essential glycolate-metabolizing enzymes. These microbodies were called "leaf peroxisomes" because enzymatically they closely resembled peroxisomes in liver and kidney cells. A detailed fine structural study of leaf peroxisomes a year later (Frederick and Newcomb, 1969b) revealed that the leaf organelles were generally larger than microbodies in nongreen tissues and were closely appressed to chloroplasts and mitochondria (Fig. 1.7), consistent with Tolbert's scheme for their involvement in the oxidation of glycolate produced in chloroplasts.

The discoveries by Beevers' and Tolbert's groups, and the subsequent ultrastructural analyses by Newcomb and associates, provided strong support for the emerging concept that peroxisomes were common plant cell organelles and had varied roles depending on the tissue type and/or state of differentiation. Other investigators working with achlorophyllous or nonfatty tissues in the late 1960s and early 1970s still considered

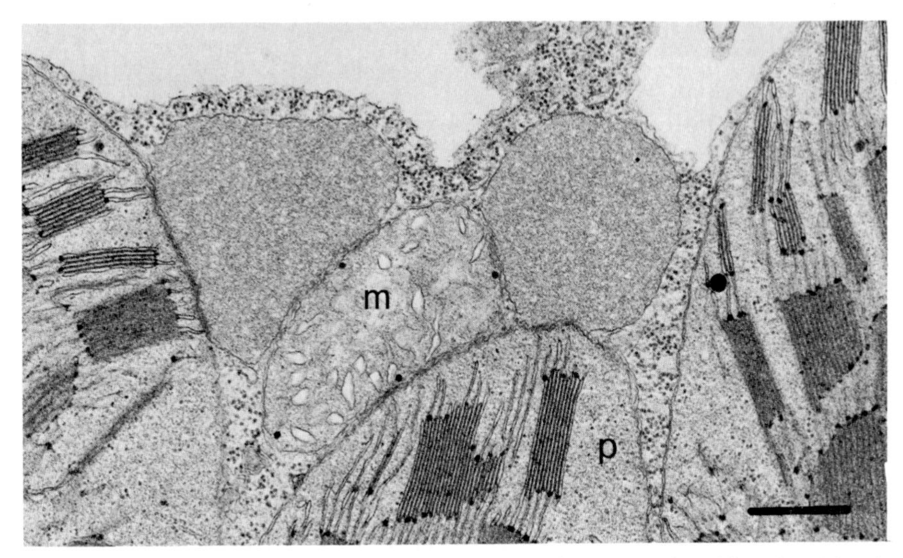

Fig. 1.7. Two microbodies (leaf peroxisomes) closely appressed to chloroplast (p) and a mitochondrion (m) in a palisade cell of a bean leaf. Bar = 0.5 μm. From P. J. Gruber (unpublished).

microbodies (or spherosomes, mostly due to confusion in terminology) as possible sites of hydrolase activity. However, a critical analysis of lysosome structure and properties in plants by Gahan (1968) stimulated plant scientists to begin looking for alternative sites. Gahan summarized "Thus, while lysosome-like particles appear to be present from optical microscope studies, all localized reactions do not appear to localize themselves as subcellular particles when viewed in the electron microscope." Low recovery of hydrolytic enzyme activity in organelle fractions indicated that plant lysosomes were not stable structures. Cytochemical studies on root tips, and biochemical and cytochemical studies with fungi, pointed to small vacuoles as the sites of hydrolytic action. In a monograph on plant lysosomes, Matile (1975) summarized much of his own and related work that demonstrated that Golgi bodies, Golgi vesicles, vacuoles, and aleurone grains (protein bodies) contained various hydrolytic enzymes. Spherosomes (oil bodies) were considered a type of lysosome only in cases where lipase was specifically associated with the oil bodies (such as in castor bean endosperm). Electron microscopic cytochemistry and cell fractionation work on microbodies in achlorophyllous and nonfatty tissues from the early 1970s to the present have shown that microbodies are not the sites of hydrolytic enzyme activity. On the contrary, microbodies in a wide range of cell types from tissues of both higher and lower plant groups have been found to contain at least catalase activity, and, where examined, an H_2O_2-producing oxidase activity. Hence, the organelles described originally in a variety of plant cells with the electron microscope, and generally called microbodies, have been characterized sufficiently in cytochemical and biochemical studies to uniformly be referred to as types of peroxisomes.

III. SUMMARY

Microbodies are a distinct class of organelles in plant cells, being recognized as such nearly 12 years after the original electron microscopic description of similar organelles in mammalian cells. Lack of recognition was due to overlapping assignment of names (and hence function) to the same organelle in different cells, and a proposed ontogenetic relationship with lipid bodies. A spherosome (a commonly misused term) is a lipid body, not ontogenetically related to microbodies. Hydrolytic enzyme activity (except for the alkaline lipase in oil seedlings) is not associated with microbodies, but is compartmented elsewhere in the cell. Microbodies in plant cells now can be regarded as peroxisomes, involved in at least H_2O_2 detoxification. They are further subcategorized depending on their known metabolic function as glyoxysomes, leaf peroxisomes, peroxisomes for other special metabolism, or unspecialized peroxisomes.

2

Cytochemistry, Morphology, and Distribution

I. GENERAL COMMENTS

The ultrastructure of peroxisomes is relatively simple. They have a single membrane surrounding a uniform matrix and lack internal membrane structure. The peroxisome matrix is composed of substances that take on a finely granular or flocculent appearance when viewed in thin sections. Inclusions of variable size and fine structure may occur within the matrix. Some of the inclusions are organized as well-ordered crystalline arrays, whereas others may appear simply as densely stained amorphous areas. Such structural features of peroxisomes are so general that it is often difficult to distinguish these organelles from other variably sized and shaped single-membrane-bound cell structures. A common feature of peroxisomes, however, is their possession of catalase. This can be readily demonstrated with the electron microscope by incubating cells or tissue slices in 3,3′-diaminobenzidine (DAB) and H_2O_2 to cytochemically localize catalase reactivity. Osmium black, formed at the site of enzyme reactivity, is a high resolution electron-dense stain, which is easily recognized in thin sections. Thus, fine structural descriptions are often accompanied by DAB cytochemistry to define the distribution and location of peroxisomes among and within cells. The cytochemistry of catalase and other peroxisomal enzymes is considered in more detail in the following section.

Documentation of peroxisome distribution among various tissues has come largely from electron microscopic investigations. Cell fractionation studies have complemented these observations, but are much less extensive in scope. Results from these combined studies have led to recognition of several types of peroxisomes: glyoxysomes, leaf peroxisomes, peroxisomes for other special metabolism, and unspecialized peroxisomes. Sections III,A and III,B describe and tabulate the morphology and distribution of these peroxisomal types in cells and tissues of the various plant groups. Specific citations to the original published work are not given in the tables. The reader may refer to a monograph by Gerhardt (1978) on plant microbodies for many of the references pertaining to the listings.

II. ENZYME CYTOCHEMISTRY

Localization of enzyme reactivity within peroxisomes via electron cytochemistry has been rather extensive when compared to cytochemistry of other plant cell organelles. This is especially true in the case of catalase localization with the DAB procedure. Examples of catalase localization within the matrix and/or inclusions of all three peroxisomal types in all groups of plants are tabulated in Sections II,A and II,B. The ubiquitous distribution of catalase activity in peroxisomes as shown by electron cytochemistry and biochemical assays of isolated organelles is the main rationale for calling plant microbodies peroxisomes. Continued use of the microbody term implies that their enzyme content is not known, which is no longer the case.

A. Catalase: 3,3′-Diaminobenzidine Oxidation

The commonly applied cytochemical procedure for localizing catalase reactivity depends on the peroxidatic oxidation of DAB by catalase and the spontaneous formation of a cyclized and polymerized form of oxidized DAB. This polymer is osmiophilic, that is, upon subsequent immersion of DAB-treated tissue segments in aqueous osmium tetroxide, the osmium radicle reacts with the oxidized DAB polymer to form "osmium black." Chelated osmium is insoluble in water and the organic solvents used for plastic embedment. Moreover, it is completely amorphous and its electron density is such that it can be observed in thin sections without poststaining. The following expressions summarize the reaction sequence of DAB:

$$DAB_{red} + H_2O_2 \xrightarrow[\text{of catalase}]{\text{peroxidatic activity}} 2\ H_2O + DAB_{ox} \quad (2.1)$$

$$DAB_{ox} \xrightarrow[\text{polymerization}]{\text{spontaneous oxidative}} \text{polymerized DAB} \quad (2.2)$$

$$\text{Polymerized DAB} + OsO_4 \rightarrow \text{Osmium black} \quad (2.3)$$

The amorphous property of the osmium black product is highly desirable for electron cytochemistry since much greater resolution of the enzyme site can be achieved when compared to granular or coalesced cytochemical products. However, a major drawback of DAB methods is the lack of specificity. The oxidation of DAB can be catalyzed by several hemoproteins, including cytochrome oxidase and peroxidase. The design of the incubation medium and several controls are important for demonstrating specificity of a particular enzyme cytochemistry. The original reaction mixture used by Novikoff and Goldfischer (1969) for localizing catalase reactivity in peroxisomes of mammalian cells has become a standard medium for localizing the enzyme in cells of higher plants (Frederick and Newcomb, 1969a; Vigil, 1973) (For examples see Figs. 2.1, 2.10, 2.15, 2.17, 2.19, and 2.21). It contains 10 mg of DAB in 5 ml of 0.05 *M* propanediol buffer containing 0.06% H_2O_2, the final pH being adjusted to 9.5. Tissue segments previously fixed in buffered 3% glutaraldehyde for at least 1 hr and washed in propanediol buffer are incubated for 30–60 min in this fresh reaction mixture. Following incubation and buffer washes, the tissue segments are postfixed in 1–2% buffered osmium tetroxide. During the latter step, osmium is chelated at the sites of catalase-produced, polymerized DAB to form the osmium black.

The specificity of the reactivity in peroxisomes is shown in several ways. The peroxidatic activity of peroxidase and cytochrome oxidase is generally optimal near neutrality, whereas the peroxidatic activity of catalase is greatest at alkaline pHs (9–10) (Vigil, 1973). Moreover, glutaraldehyde fixation inhibits the catalatic activity of catalase while enhancing the peroxidatic activity of catalase (Herzog and Fahimi, 1974). Several controls are done with the standard procedure which include (1) adding aminotriazole to the incubation medium as a potent inhibitor of all catalase activity, (2) adding potassium cyanide to the medium to inhibit peroxidase activity and the peroxidatic activity of catalase, and/or (3) omitting H_2O_2 from the medium. Peroxisomes often are stained when H_2O_2 is absent from the medium, but this has been attributed to endogenous H_2O_2 production by oxidases in the peroxisomes (Vigil, 1973). Finally, the large body of biochemical evidence for catalase activity in isolated peroxisomes and glyoxysomes has reinforced the cytochemical demonstration of catalase reactivity. Reaction product observed in mitochondrial cristae is attributed to cytochrome oxidase, and any vacuolar and cytosolic depositions are considered due to peroxidases. Examples of catalase reactivity in peroxisomes are shown in Figs. 2.1, 2.10, 2.15, 2.17, 2.19, and 2.21. Osmium black is distributed throughout the peroxisomal matrices, and when crystalloids are present, the reaction product is found in both the crystalloids and in the surrounding matrix (Fig. 2.1).

B. Malate Synthase: Ferricyanide Reduction

The cytochemistry of malate synthase, a marker enzyme of the glyoxylate cycle, originally was developed to detect the enzyme in glyoxysomes of germinated oilseeds (Trelease *et al.*, 1974). Catalase cytochemistry is useful for identifying peroxisomes among other cellular organelles, but cannot be used to distinguish glyoxysomes from other peroxisomes.

The cytochemical localization of malate synthase is based on the reduction of ferricyanide by free coenzyme A (CoA) produced by the malate synthase reaction. The ferrocyanide then reacts immediately with cupric ions to yield copper ferrocyanide (also called Hatchett's brown) which is an insoluble electron-dense product. This product can also be observed with the light microscope as a brownish-red deposit. The general reaction scheme is as follows:

$$\text{Acetyl-S-CoA} + \text{glyoxylate} \xrightarrow[\text{malate synthase}]{Mg^{2+}} \text{L-malate} + \text{CoA-SH} \quad (2.4)$$

$$2\ \text{CoA-SH} + 2\ \text{ferricyanide} \rightarrow 2\ \text{ferrocyanide} + \text{CoA-S-S-CoA} + 2\ H^+ \quad (2.5)$$

Fig. 2.1. (Top) Portion of a tobacco leaf cell fixed and poststained in the usual manner (i.e., without incubation in DAB medium). A peroxisome with a large crystalline inclusion is appressed to two chloroplasts (C). A small mitochondrion (m) lies to the right of the peroxisome. (Bottom) Portion of a tobacco leaf cell incubated in the DAB medium. The peroxisome contains a large crystalline inclusion which shows a heavy deposition of osmium black throughout. Reaction product also is visible in the peroxisome matrix around the crystal. There is no product in the chloroplasts (C) or in the mitochondrion (m). Bars = 0.5 μm. [From Frederick and Newcomb. *J. Cell Biol.* (1969) **43**, 343–353, by permission of Rockefeller University Press.]

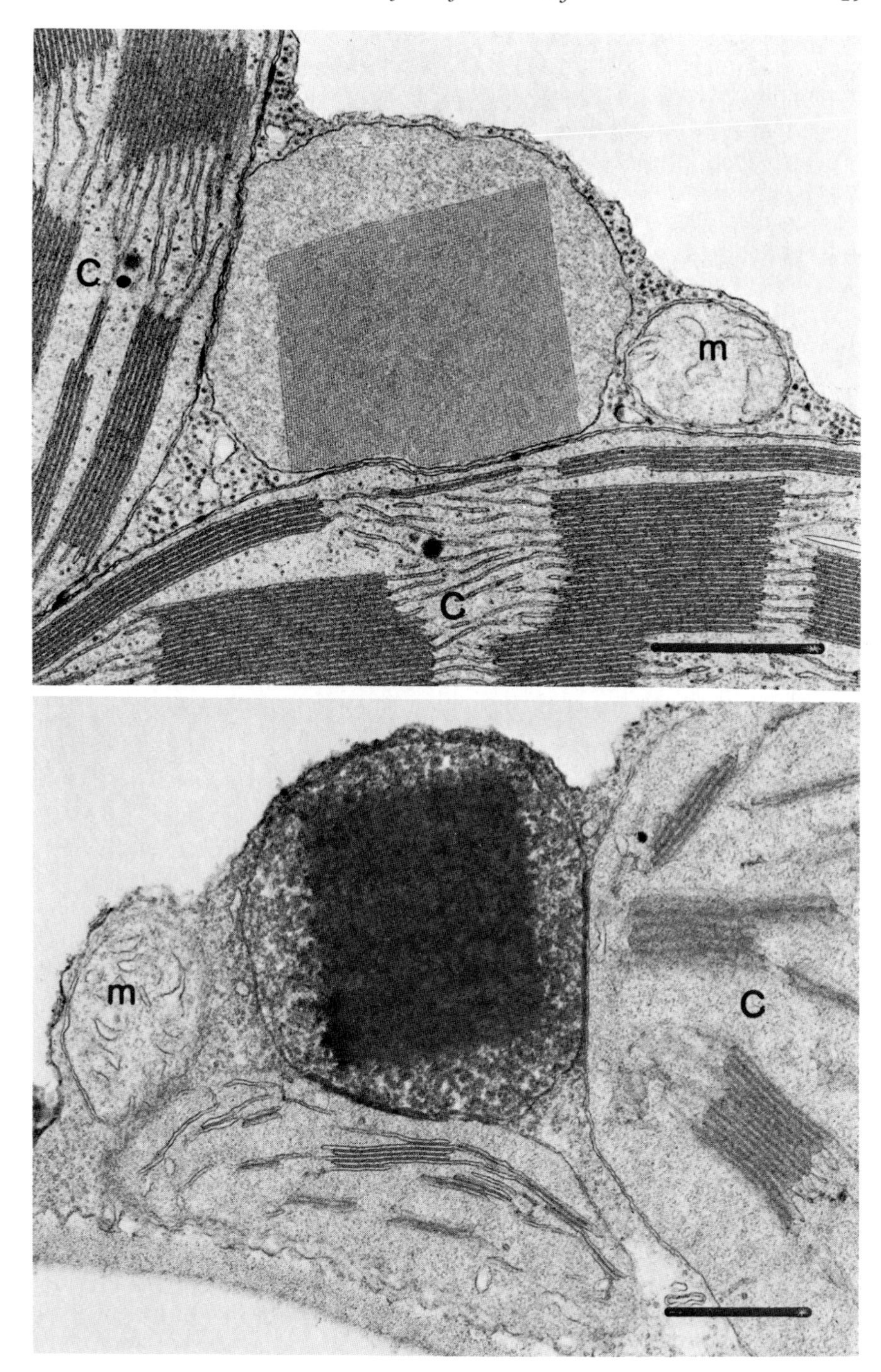
C
m
C
m
C

$$\text{Ferrocyanide} + \text{cupric ion} \rightarrow \text{copper ferrocyanide (Hatchett's brown)} \quad (2.6)$$

The standard reaction mixture consists of the following components added in the indicated order with stirring between each addition (from Trelease, 1975).

Amount (ml)	Stock solution	Final concentration (m*M*)
0.30	65 m*M* K-phosphate (pH 6.9)	19.5
0.20	50 m*M* copper sulfate–500 m*M* Na,K-tartrate solution (pH 6.9)	10, 100
0.25	Distilled H_2O	—
0.03	50 m*M* potassium ferricyanide	1.5
0.10	50 m*M* magnesium chloride	5.0
0.02	150 m*M* Na-glyoxylate	3.0
0.10	10 m*M* acetyl-CoA	1.0

Tissue segments from one to two oilseed cotyledons are fixed in buffered 4% formaldehyde–1% glutaraldehyde for a short period. The objective is to obtain satisfactory ultrastructural preservation of cellular components without inactivating the malate synthase. Following incubation in the above mixture for 30–60 min, the segments are washed, postfixed in OsO_4, and embedded in plastic in the usual manner. Reaction product is deposited uniformly in the matrix of the glyoxysomes (Fig. 2.2). Thus far, this cytochemistry in plants has been applied only to cotyledons of certain oilseeds (cucumber, cotton, sunflower) (cucumber in Fig. 2.2, left), zoospores of the fungus *Entophlyctis* sp. (Fig. 2.2, right) and germinated spores of the moss *Bryum cappilare* (Pais and Carrapico, 1979). Recently, malate synthase activity was shown present by this method in peroxisomes of the toad urinary bladder epithelium (Goodman *et al.*, 1980; Jones *et al.*, 1981). The method is specific for malate synthase reactivity, but a significant limitation is the inefficient penetration of reactants into cells. This limitation most likely explains the lack of a more widespread application of the technique. Nevertheless, such a specific marker for glyoxylate cycle activity can be extremely useful for discovering the function of microbodies in tissues where few biochemical data currently are available. It would be particularly important for analyzing tissues that are difficult to obtain sufficiently large quantities for cell fractionation experiments, (i.e., toad bladder epithelium and *Bryum cappilare* spores).

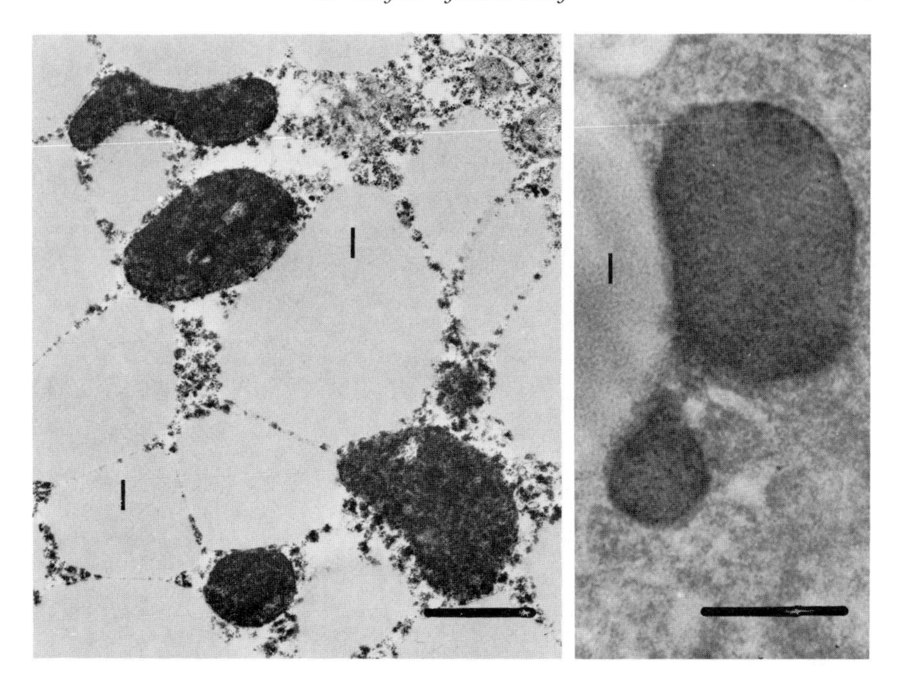

Fig. 2.2. Glyoxysomes with electron-dense deposits attributable to malate synthase reactivity following incubation in a copper ferricyanide reaction mixture. (Left) Glyoxysomes among lipid bodies (l) in a 4-day-old cucumber cotyledon cell showing reaction product distributed throughout the glyoxysomal matrices. Bar = 1 μm. From R. N. Trelease (unpublished). (Right) Appearance of reacted peroxisomes adjacent to a lipid body (l) in a zoospore of *Entophlyctis* sp. The reaction product is evenly distributed in the matrix of the peroxisomes as in the cucumber cotyledon cells. This section was not poststained. Bar = 0.5 μm. From Powell (1976).

C. Glycolate Oxidase: Ferricyanide Reduction

The cytochemistry of flavin-linked glycolate oxidase, a marker enzyme for leaf peroxisomes (although found in low levels in other peroxisomes), was developed originally using ferricyanide reduction as a method for detecting α-hydroxyacid oxidase reactivity in mammalian cells (Shnitka and Talibi, 1971). The first published application of this cytochemistry to plant material was done on cotyledon cells of greening cucumber seedlings (Burke and Trelease, 1975). To date, this procedure has been extended only to localizing glycolate and other α-hydroxyacid oxidases in a green alga, *Klebsormidium* (Gruber and Frederick, 1977), and *Bryum cappilare* spores (Pais and Carrapico, 1979). Again, the lack of a more widespread application may be due to its inherent limitations as a ferricyanide-reduction type of cytochemistry. The reaction scheme for the glycolate oxidase cytochemistry via ferricyanide reduction is as follows:

$$\text{Glycolate} + \text{PMS}_{ox} \xrightarrow[\text{oxidase}]{\text{glycolate}} \text{glyoxylate} + \text{PMS}_{red} + \text{H}^+ \quad (2.7)$$

$$\text{PMS}_{red} + \text{ferricyanide} \rightarrow \text{ferrocyanide} + \text{PMS}_{ox} \quad (2.8)$$

$$\text{Ferrocyanide} + \text{cupric ion} \rightarrow \text{copper ferrocyanide (Hatchett's brown)} \quad (2.9)$$

The standard reaction mixture consists of the following components added in the indicated order with stirring between each addition (from Burke and Trelease, 1975):

Amount	Stock solution	Final conc. (m*M*)
8.5 ml	25 m*M* Potassium phosphate (pH 7.2)	21
0.5 ml	60 m*M* Copper–40 m*M* tartrate solution (pH 7.2)	3, 2
1.0 ml	15 m*M* potassium ferricyanide	1.5
5 mg	Phenazine methosulfate (PMS)	1.6
2 mg	Flavin mononucleotide	0.4
12.5 mg	Sodium glycolate	12.8

Tissue segments from one to two cotyledons or a comparable amount of tissue are fixed as for malate synthase cytochemistry; glycolate oxidase also appears to be sensitive to excessive fixation by glutaraldehyde. It is not necessary to incubate the segments anaerobically. Apparently both oxygen and phenazine methosulfate can accept electrons from glycolate.

Deposition of copper ferrocyanide is generally visualized throughout the peroxisomal matrices (Figs. 2.3 and 2.4, left), but a peculiar peripheral staining also is seen frequently (Figs. 2.3 and 2.4, right). Peroxisomes in the same cell (often adjacent) may show both types of staining. Assuming that the same enzyme generates these patterns, the type of substrate may have some influence. Peripheral staining is more prevalent with the substrate most actively oxidized by a particular α-hydroxyacid oxidase (e.g., glycolate and not lactate for glycolate oxidase). This has led to the suggestion that the "halo" pattern is a consequence of product diffusion prior to capture by the cupric ions. Direct evidence for this proposal is lacking. It also has been suggested that the alternate staining patterns reveal differential localizations of the enzyme within

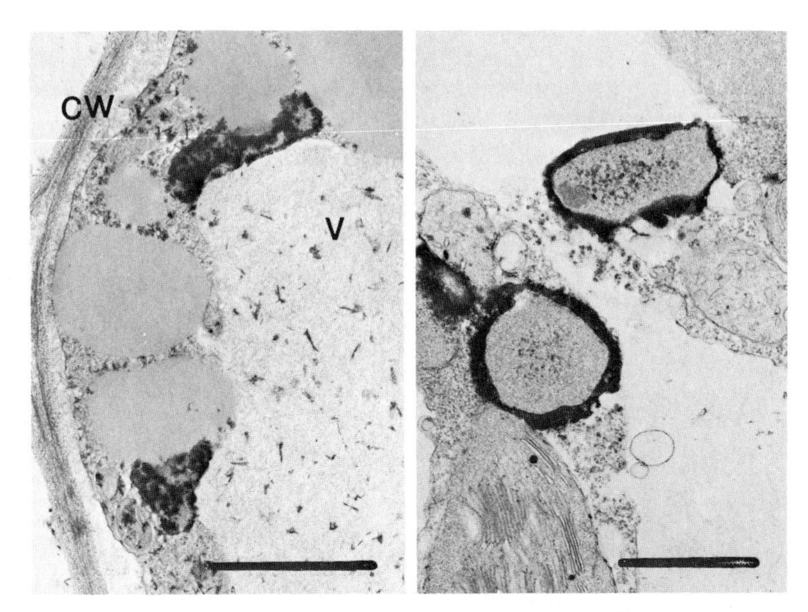

Fig. 2.3. Cytochemical localization of glycolate oxidase in leaf-type peroxisomes of green cucumber cotyledons (5-day-old). (Left) Cells incubated in ferricyanide media with glycolate as substrate yielding copper ferrocyanide deposition in the peroxisomal matrices. CW, cell wall; V, vacuole. Bar = 1 μm. From Burke and Trelease (1975). (Right) Adjacent cell incubated under same conditions as in left micrograph, but glycolate oxidase reaction product is observed surrounding the peroxisomes, not uniformly distributed within their matrices. Bar = 1 μm. From J. J. Burke (unpublished).

the organelle, but recent results with cerium perhydroxide precipitation make this proposal less tenable (see Section II,D).

D. α-Hydroxyacid, Alcohol, Amine, and D-Amino Acid Oxidases: Cerium Perhydroxide Precipitation

A new technique employing cerium as the metal in the electron-dense product has been used successfully to demonstrate the peroxisomal localization of several different oxidases in methanol-grown yeast cells (Veenhuis *et al.*, 1976) and glycolate oxidase in peroxisomes of higher plant tissues (Thomas and Trelease, 1981). The method was first introduced by Briggs *et al.* (1975) to localize an NADH oxidase in mammalian polymorphonuclear leukocytes. The principle of the cytochemistry is to have cerium ions react with the hydrogen peroxide which is enzymatically produced by the oxidase. The produce is cerium perhydroxide, which presumably occurs in at least two forms, produced by two separate reactions [Eq. (2.11) or (2.12)]. Cerium perhydroxide precipi-

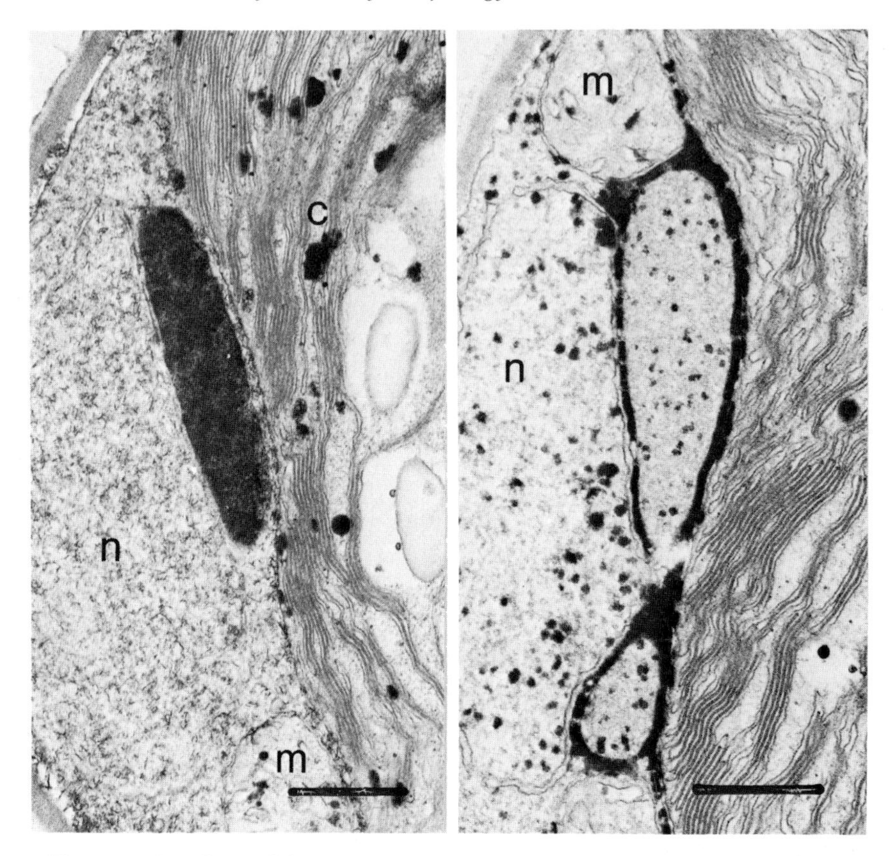

Fig. 2.4. Cytochemical localization of glycolate oxidase in peroxisomes of *Klebsormidium flaccidum* filaments. Left—Electron-dense stain (Hatchett's brown) entirely fills the peroxisome while only a few background deposits are present in the chloroplast (c), mitochondrion (m), and nucleus (n). Bar = 0.5 μm. (Right) Cell incubated in a similar reaction mixture, but the stain surrounds the periphery of the peroxisome rather than occurring in the matrix. m, mitochondrion; n, nucleus. Bar = 0.5 μm. From Gruber and Frederick (1977).

tates have properties highly suitable for electron cytochemistry: the product is less granular than copper ferrocyanide, thus allowing greater resolution of the enzyme site (compare Figs. 2.3 through 2.9). The presumed reaction sequences for product formation are as follows:

$$\underset{\text{reduced substrate}}{RH_2} + O_2 \xrightarrow{\text{oxidase}} \underset{\text{oxidized product}}{R} + H_2O_2 \quad (2.10)$$

$$3\ CeCl_3 + 3/2\ H_2O_2 + 12\ H_2O \longrightarrow \underset{\text{cerium perhydroxide}}{3\ Ce(OH)_3OOH} + 9\ HCl + 6\ H^+ \quad (2.11)$$

and/or

$$CeCl_3 + H_2O_2 + 2\,H_2O \longrightarrow \underset{\text{cerium perhydroxide}}{Ce(OH)_2OOH} + 3\,HCl \tag{2.12}$$

The cytochemical procedure outlined below is basically that of Veenhuis *et al.* (1976), which is slightly modified from Briggs *et al.*'s (1975) original procedure. Either unfixed or glutaraldehyde-fixed cells (or tissue segments) are preincubated in 100 m*M* Tris–maleate, pH 7.5, containing 50 m*M* aminotriazole (to inhibit endogenous catalase activity) and 5 m*M* $CeCl_3$ for 30 min, then incubated in the same medium supplemented with 50 m*M* substrate (e.g., sodium urate, methanol, D-alanine, sodium glycolate, or sodium DL-lactate). The complete reaction mixtures are aerated continuously during a 3-hr incubation. Thereafter, the material is washed in 100 m*M* cacodylate buffer, pH 6.0, for 30 min, to solubilize any nonspecific cerium hydroxide precipitate that may form during incubation. Postfixation in 1% OsO_4 and 2.5% $K_2Cr_2O_7$ in 0.1 *M* cacodylate buffer was done with the yeast cells (Figs. 2.5 and 2.6).

Veenhuis *et al.* (1976) noted that the pH of the incubation medium with methanol as a substrate decreased from 8.5 to 7.9, or from 7.5 to 6.5, within 30 min of incubation. They regularly checked the pH, and renewed the incubation medium every hour during the 3-hr incubation period to help overcome this pH problem, and to remove any cerium hydroxide formed during aeration. They attributed the decrease in pH in methanol-containing media to the incomplete oxidation of methanol to formic acid. However, if one examines the reaction sequences given above, it is apparent that substantial HCl is produced as a by-product of cerium perhydroxide deposition. This also may contribute to the decreased pH.

The methanol oxidase reactivity in yeast cells was detected only in crystal-containing peroxisomes. Veenhuis *et al.* interpreted their results as showing enzyme reactivity in both the crystalloid proper and the noncrystalloid matrix of these peroxisomes (Fig. 2.5, top). D-Alanine, urate, glycolate, and DL-lactate oxidase activities also were localized in peroxisomes and showed a similar staining pattern to that of methanol oxidase (Fig. 2.5 middle and bottom, and Fig. 2.6). Reactivities of the latter four oxidases, however, also occurred in non-crystal-containing peroxisomes, distributed throughout the matrix (Veenhuis *et al.*, 1979). Details of the biochemical aspects of methanol- and *n*-alkane-grown yeast cells are given in Chapter 4.

The method was adapted by Thomas and Trelease (1981) for the cytochemical localization of glycolate oxidase in leaf peroxisomes (guayule and tobacco leaves; green cucumber cotyledons) (e.g., Fig.

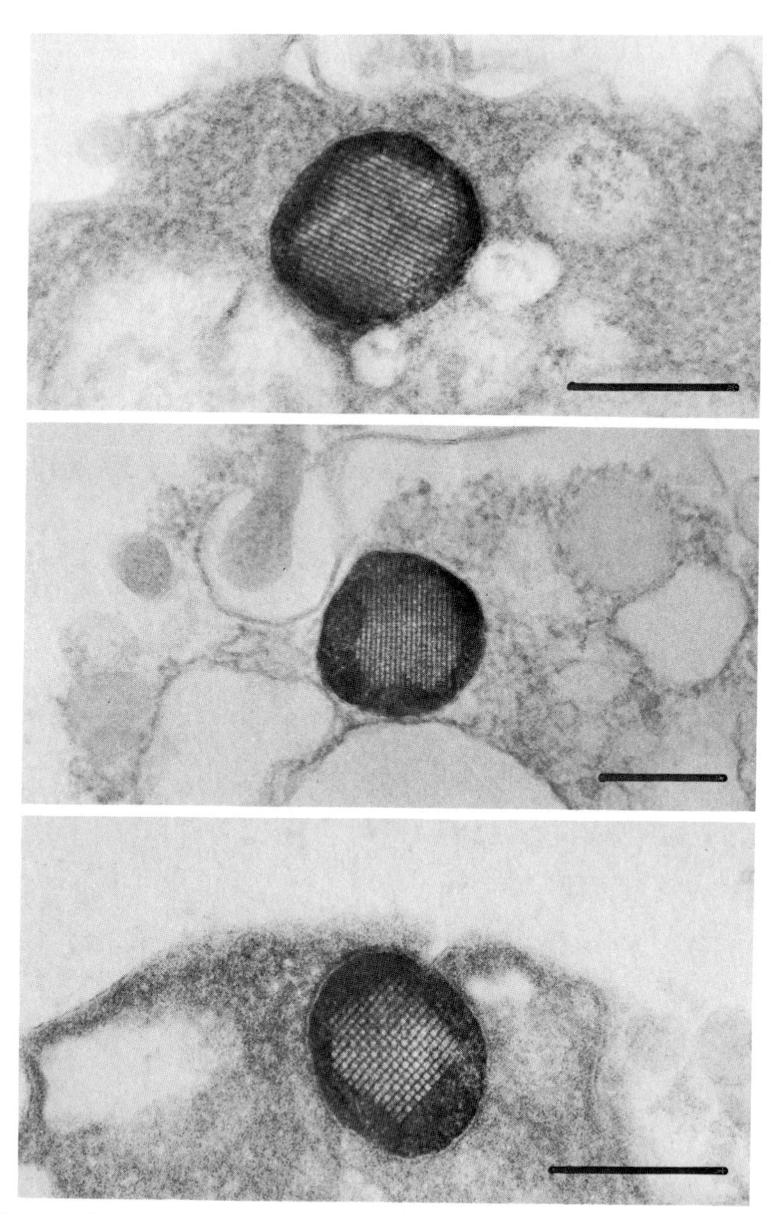

Fig. 2.5. Peroxisomes in *Hansenula polymorpha* spheroplasts 6 hr after transfer into a methanol medium from a glucose medium with different substrates and postfixed in $OsO_4/K_2Cr_2O_7$. All peroxisomes show reaction product in the matrix surrounding the crystalloids which are clearly defined. (Top) Alcohol oxidase reactivity with methanol as substrate. (Middle) D-Amino oxidase reactivity with D-alanine as substrate. (Bottom) α-Hydroxyacid oxidase reactivity with DL-lactate as substrate. Bars = 0.25 μm. From Veenhuis *et al.* (1979).

2.7), glyoxysomes (castor bean and cucumber) (e.g., Fig. 2.8), and unspecialized peroxisomes (etiolated barley leaf and barley coleoptile (Fig. 2.9). The method employed was modified only slightly from that of Veenhuis *et al.* (1976). Rather than continuously aerating the complete reaction mixture containing tissue segments, they aerated the solutions prior to incubation, but changed the mixture every hour during a 3-hr incubation. For those tissues containing relatively high glycolate oxidase activity (i.e., green tissues), the 3-hr incubation was satisfactory. However, to reproducibly detect the enzyme in glyoxysomes and unspecialized peroxisomes, they had to extend the incubation period in the complete reaction mixture to 18 hr. Solutions again were aerated prior to adding tissue segments; the mixture was changed only once during the 18 hr (after 3 hr). Some minor globular depositions occurred in the cells incubated for 18 hr (Figs. 2.8 and 2.9), but did not detract from the enzyme localization in the peroxisomes.

In all types of peroxisomes examined, glycolate and other oxidases have been localized in the matrix. However, there is disagreement on whether the crystalloids and other inclusions in the higher plant and yeast peroxisomes contain these enzymes. Veenhuis *et al.* (1976, 1979) interpret their micrographs as showing that alcohol and other oxidases

Fig. 2.6. Glycolate oxidase localization in peroxisomes of *Hansenula polymorpha* spheroplasts from a late stationary batch culture. Staining with $CeCl_3$ and glycolate. Reaction product is throughout the peroxisome, but the crystal is still visible. Bar = 0.2 μm. From Veenhuis *et al.* (1978).

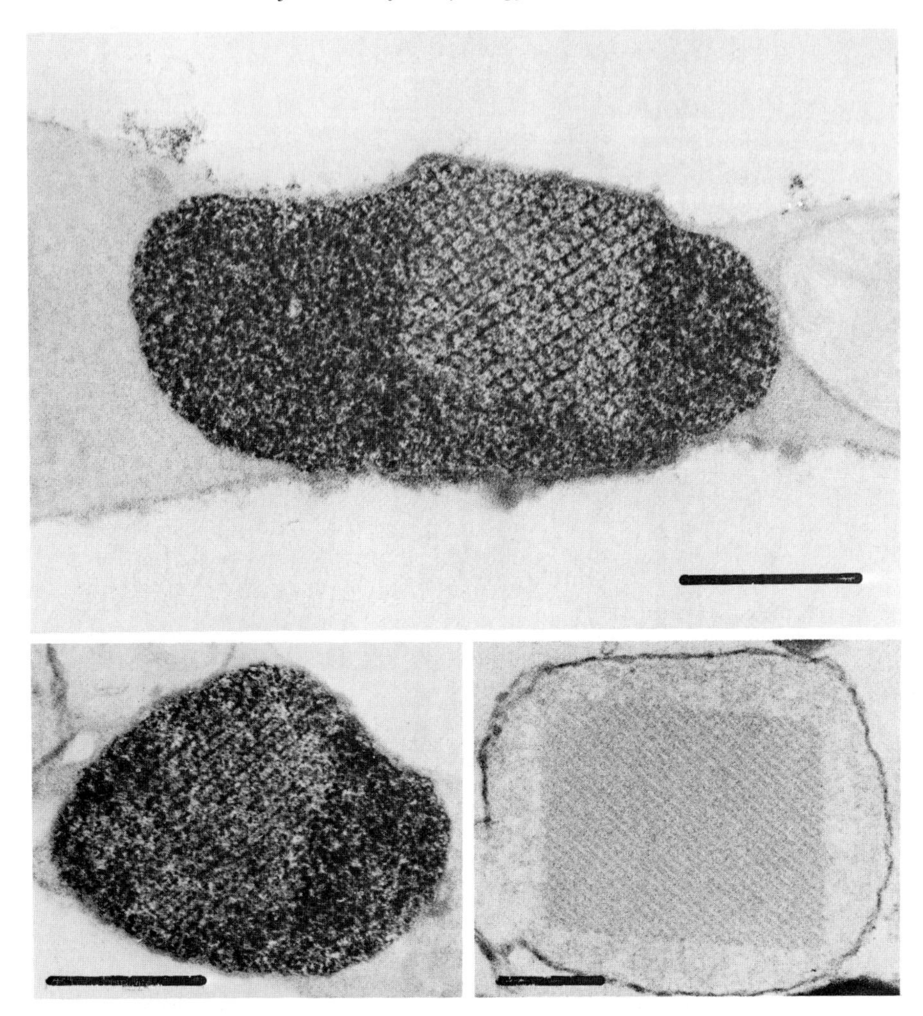

Fig. 2.7. Crystal-containing peroxisomes in guayule leaf cells. (Top and bottom left) Leaf segments were incubated with glycolate and $CeCl_3$ for 3 hr following a 30-min aldehyde fixation. Reaction product is throughout the peroxisomal matrices. Note the non-uniform depositions in the crystal areas sectioned at different planes in the two micrographs. The structural units of the crystals are not stained, whereas the spaces between them are. (Bottom right) Peroxisome in a leaf segment incubated in $CeCl_3$ for 3 hr without glycolate. Reaction product is not apparent in the matrix or crystal area. This sectional view illustrates the continuity between the matrix and spaces between the structural units of the crystal. Bars = 0.25 μm. From Thomas and Trelease (1981).

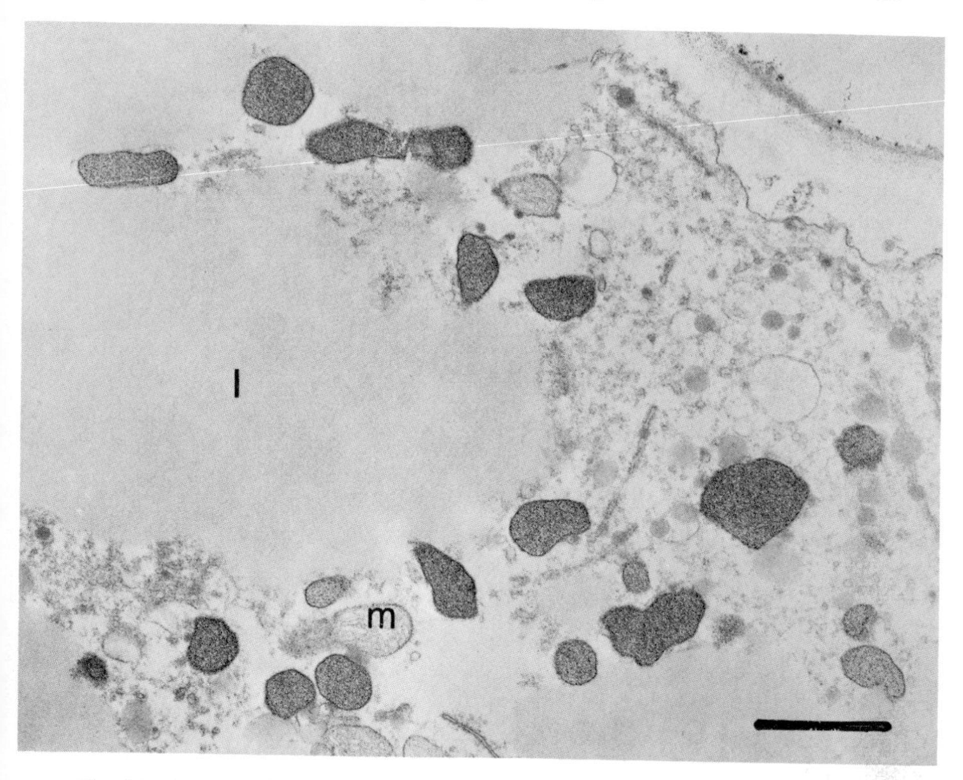

Fig. 2.8. Portion of a castor bean endosperm cell from a segment fixed for 10 min and incubated with glycolate and $CeCl_3$ for 18 hr. All the glyoxysomes show deposition of reaction product in their matrices. Globular background deposits are associated with ER, mitochondria (m), and vesicles. l, lipid body. Bar = 0.5 μm. From Thomas and Trelease (1981).

are a major component of the crystalline inclusions. Their conclusions also are based on correlations between increases in oxidase activities and the appearance of crystal-containing peroxisomes. Thomas and Trelease (1981) contend that the structural units proper of the crystalloids are not stained for glycolate oxidase but that the spaces *between* the structural units accumulate cerium perhydroxide depositions. They came to this conclusion by comparing unstained control preparations with stained peroxisomes (see Fig. 2.7), and presumed that if the crystals were stained, the crystalloids would appear as those showing catalase staining of crystalloids (Fig. 2.1). It is an interesting point because the staining patterns obtained by the two groups essentially are the same (cf. Figs. 2.6 and 2.7). Each group simply has interpreted the staining patterns differently. That glycolate oxidase is not part of the crystalline

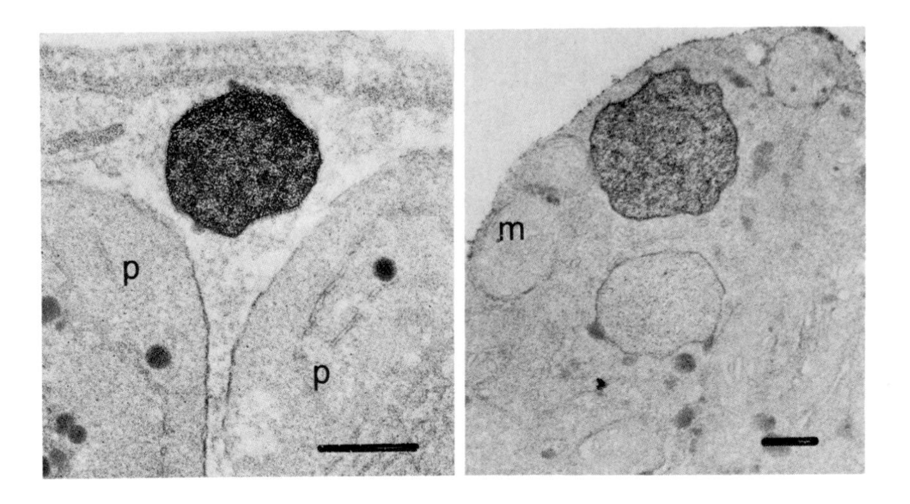

Fig. 2.9. Portion of cells in an etiolated barley leaf (left) and barley coleoptile (right) showing unspecialized peroxisomes stained for glycolate oxidase activity. The tissue segments were incubated in a glycolate and $CeCl_3$ for 18 hr following a 15-min aldehyde fixation. Peroxisomes in both preparations show reaction product in the matrix, although staining is much lighter in the coleoptile peroxisome. p, plastid; m, mitochondrion. Bars = 0.25 μm. From Thomas and Trelease (1981).

inclusions was not proved; the possibility remained that if the enzyme were incorporated into a structural unit of a crystal, it could be cytochemically unreactive (Thomas and Trelease, 1981). In summary, the available cytochemical data show that oxidases are localized in the matrices of the various types of peroxisomes, but it remains unresolved as to whether the oxidases are incorporated into the structural units of peroxisomal inclusions. Biochemical studies on isolated crystals have not been done to help resolve this question.

The $CeCl_3$ method is relatively simple and apparently adaptable for localizing several different oxidase activities. The resolution of the cerium perhydroxide deposits is comparatively good, and the method does not yield the bothersome nonspecific background depositions commonly associated with ferricyanide-reduction cytochemistry. Thus far, the technique has been used to localize several oxidases in yeast and mammalian cells, and only glycolate oxidase in higher plant cells. Attempts were made to localize urate oxidase and fatty acyl-CoA oxidase in higher plant peroxisomes (Thomas and Trelease, 1981), but were not successful, apparently due to relatively low activities in the cells tested and pH problems. The success already achieved and the high resolution in the localization of the reaction product indicate that further work on modifying this procedure should be done to provide valuable information on oxidase localization in the various types of peroxisomes.

III. MORPHOLOGY AND DISTRIBUTION

A. Angiosperms

1. Achlorophyllous and Nonoil Storage Tissues

These tissues usually contain the so-called unspecialized peroxisomes and are found in all living tissues in flowering plants. Electron microscopic examinations of angiosperm tissues have shown that peroxisomes are constituent organelles of their cells (Table 2.1). Although the list of plant parts is reasonably comprehensive, the number of reports per plant part, with the exception of roots, is rather limited. Cytochemistry of catalase within unspecialized peroxisomes (e.g., Fig. 2.10) has been demonstrated only in approximately one-third of the listings (Table 2.1). Nevertheless, the widespread distribution of unspecialized peroxisomes within various plant parts and species is clearly indicated.

Peroxisomes in these tissues generally are smaller in diameter and occur less frequently in cells than do glyoxysomes or leaf peroxisomes. Specific associations with other organelles are seldom seen, except for a common association with profiles of rough endoplasmic reticulum (Figs. 1.1, 1.3, 1.5, 2.11, 2.12, and 2.20). The organelles tend to have little contrast between their matrix and the surrounding cytosol, making direct observation difficult (Fig. 1.3). Their shapes can vary from nearly spherical to ovoid (Figs. 1.3 and 2.11), or they may assume a dumbbell appearance in profile view (Figs. 1.3 and 2.12). Inclusions within their matrices are also variable. For example, a dense amorphous nucleoid occupies a small portion of the peroxisome matrix in a pumpkin hypocotyl (Fig. 2.11), whereas a well-developed crystal occupies nearly the entire peroxisome interior in an oat coleoptile (Fig. 1.5).

Isolation of unspecialized peroxisomes and assay for enzyme activities provide further evidence that they contain catalase (Table 2.2). They generally equilibrate at a lower buoyant density (1.22–1.24 g/cm^3) on sucrose gradients than glyoxysomes and leaf peroxisomes (Huang and Beevers, 1971; Gerhardt, 1978). They are usually more difficult to isolate intact due to having fewer organelles per cell, the toxic compounds released from the large cell vacuoles during homogenization, etc. Besides catalase, glycolate oxidase and urate oxidase are the only known

TABLE 2.1

Achlorophyllous and Nonoil Storage Tissues of Angiosperm Species Shown to Possess Microbodies by Electron Microscopic Examination of Thin Sections[a]

Plant part and species	DAB
Callus (solid)	
Nicotiana tabacum	
Coleoptile	
Avena sativa	+
Triticum aestivum	
Coleorhiza	
Zea mays	
Fruit	
Pyrus sp.	
Lycopersicum esculentum	
Hypocotyl	
Cucurbita pepo (pumpkin)	
Phaseolus mungo	
Latex vessel	
Euphorbia characias	+
Leaf (nongreen area or etiolated)	
Allium cepa	+
Avena sativa	+
Coleus blumei	+
Phaseolus vulgaris	+
Nicotiana tabacum (albino)	+
Nectary	
Prunus sp.	
Nodule	
Glycine max	+
Ovule	
Convolvulus sp.	
Pedicel	
Lycopersicum esculentum	+
Nicotiana tabacum	+
Phloem	
Acer rubrum	
Citrus sp. (lime)	
Pollen	
Zea mays	
Rhizoid	
Lemna minor	
Root	
Allium cepa	+
Allium sativum	+
Cattleya sp.	
Cucumis sativus	
Eichhornia crassipes	
Euphorbia characias	+
Fraxinus excelsior	
Gossypium hirsutum	+
Nicotiana tobacum	+
Phaseolus mungo	
Phaseolus vulgaris	
Phleum pratense	
Quercus sp.	
Raphanus sativus	
Ricinus communis	
Rorippa sp.	
Triticum aestivum	
Trapaeoleum majus	
Vanilla planifolia	
Zea mays	
Zingiber officinale	+
Shoot apex	
Euphorbia characias	+
Zea mays	
Spadix appendix	
Arum maculatum	+
Sauromatum guttatum	+
Stem	
Acer rubrum	
Nicotiana tabacum	
Stigma	
Convolvulus sp.	
Suspension cell	
Acer pseudoplatanus	+
Eucalyptus camaldulensis	
Tuber	
Solanum tuberosum	
Xylem	
Avena sativa	

[a] Species with peroxisomes showing cytochemical catalase reactivity by the 3,3′-diaminobenzidine (DAB) procedure are listed with a + symbol.

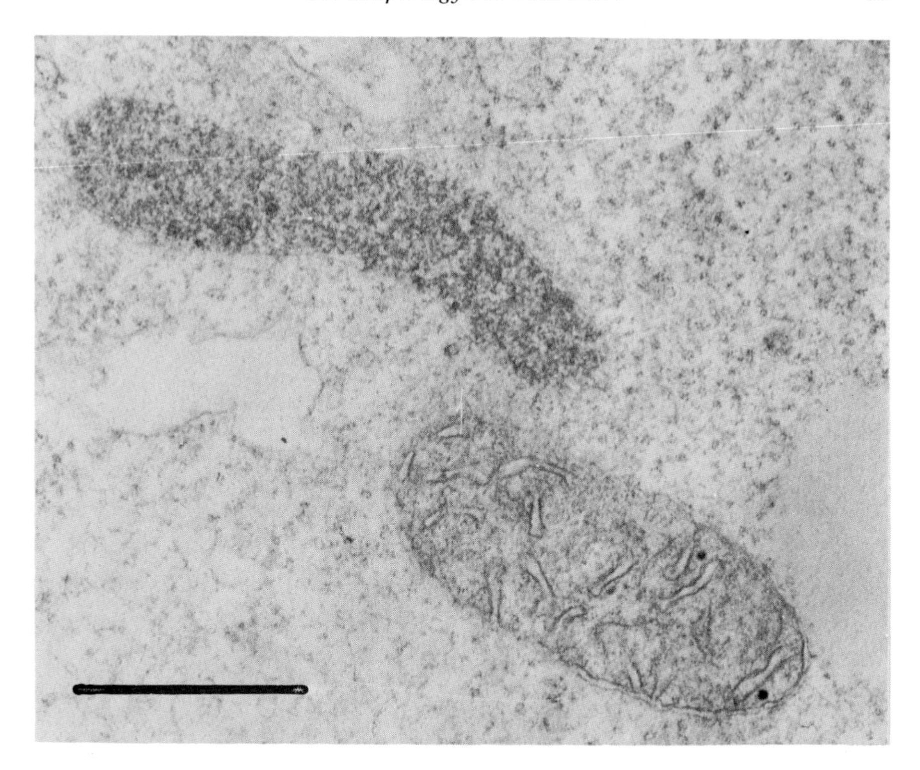

Fig. 2.10. Unspecialized peroxisome in an epidermal cell of an onion bulb scale. The tissue was treated for catalase reactivity with DAB and H_2O_2. The osmium black reaction product is throughout the peroxisome matrix. Sections were not poststained. Bar = 0.5 μm. From E. L. Vigil (unpublished).

constituent enzymes of these particles (Fig. 2.9; Table 2.2). Their general function in H_2O_2 detoxification is assumed, but this does not preclude their role in some other as yet undetermined metabolic process. An example of an apparent specialization of an unspecialized peroxisome recently was found in nodules of soybean roots (Newcomb and Tandon, 1981). Peroxisomes in the cortical root cells not infected by *Rhizobium* are considerably larger than peroxisomes in similar root cells not nodulated. It is believed that the enlarged root peroxisomes have become specialized and involved in ureide metabolism (Chapter 4).

2. Oil-Storage Tissues

The occurrence of glyoxylate cycle and fatty acid β-oxidation enzymes within the glyoxysomes distinguishes them as specialized peroxisomes playing a major role in the mobilization of triacylglycerols in oil-rich

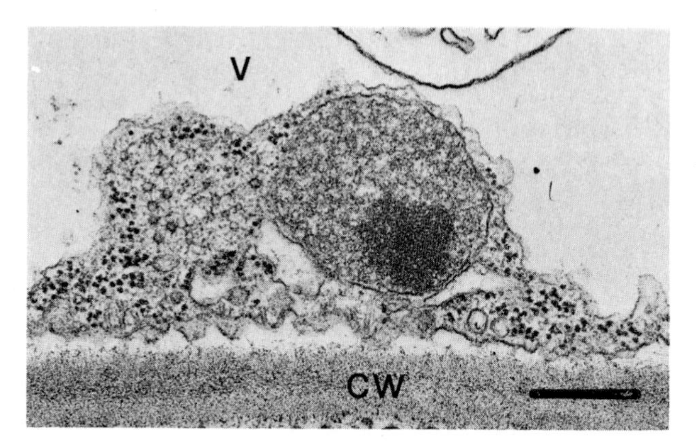

Fig. 2.11. Unspecialized peroxisome in the cytoplasm of a vacuolated parenchyma cell of a pumpkin hypocotyl. An amorphous inclusion is apparent in this relatively small organelle. V, vacuole; CW, cell wall. Bar = 0.25 μm. From R. N. Trelease (unpublished).

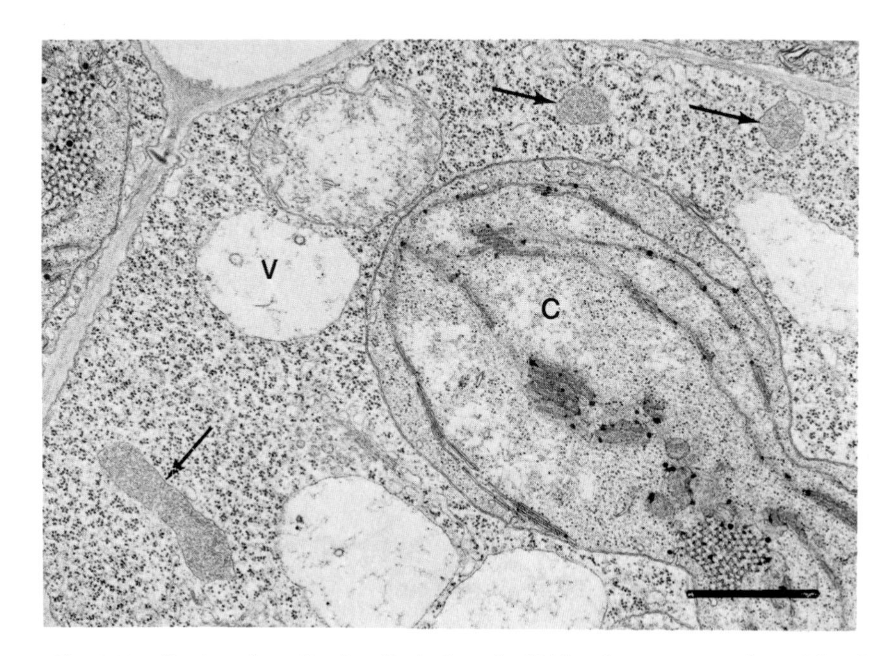

Fig. 2.12. Portion of a palisade cell of a bean leaf 24 hr after exposure of an etiolated plant to light. Three relatively small peroxisomes (arrows) are distributed in the cytoplasm not closely associated with the developing chloroplast (C). V, vacuole. Bar = 1 μm. From P. J. Gruber (unpublished).

TABLE 2.2

Achlorophyllous and Nonoil Storage Tissues of Angiosperm Species from Which Unspecialized Peroxisomes Have Been Isolated by Density Gradient Centrifugation[a]

Plant part and species	Enzyme activity		
	Catalase	Urate oxidase	Glycolate oxidase
Bud			
Brassica oleracea (cauliflower)	+	+	+
Coleoptile			
Zea mays	+	+	+
Cotyledon			
Pisum sativum	+	+	+
Spinacia oleracea	+		+
Flower petal			
Dahlia pinnata	+	–	+
Fruit			
Malus sylvestris	+	+	+
Fruit coat			
Ricinus communis	+	+	+
Hypocotyl			
Ricinus communis	+	+	+
Spinacia oleracea	+		–
Leaf (white or etiolated)			
Hedera helix (white)	+	–	+
Hordeum vulgare	+		+
Nodule			
Glycine max	+	+	
Root			
Daucus carota	+	+	+
Ipomoea batatas	+	+	–
Pisum sativum	+		–
Ricinus communis	+	+	+
Spinacia oleracea	+		–
Triticum sp.	+		
Zea mays	+	+	+
Spadix appendix			
Arum italicum	+	+	+
Arum maculatum	+	+	+
Sauromatum guttatum	+	+	+
Suspension cell			
Glycine max	+		+
Rosa (Paul's scarlet)	+		
Tuber			
Solanum tuberosum	+	+	+

[a] Enzyme activities associated with the peroxisomes are designed by +; the – symbol indicates the activity was not detected.

tissues. The major research efforts on their function have been with castor bean endosperm and cotyledons of various species (Table 2.3). Glyoxysomes are numerous in cells of these tissues, characteristically located among the storage lipid bodies (Figs. 2.2, 2.8, 2.13 through 2.16, and 2.21). Their shape seems to depend on their placement among the packed lipid bodies; this is particularly evident in the micrograph il-

TABLE 2.3

Distribution of Glyoxysomes in Oil-Storage Tissues of Angiosperms[a]

Plant part and species	EM morphology	DAB	Isolated
Aleurone			
Hordeum vulgare	X		−
Triticum vulgare	X		
Cotyledon			
Arachis hypogaea	X	+	+
Brassica napus	X	+	+
Carthamus tinctorius			+
Citrullus vulgaris	X	+	+
Cucumis sativus	X	+	+
Cucurbita pepo (pumpkin)	X		+
Cucurbita pepo (marrow)	X		+
Fraxinus americana			+
Glycine max	X		+
Gossypium hirsutum	X	+	+
Helianthus annuus	X	+	+
Lepidium virginicum	X		
Linum usitatissimum	X		
Lycopersicum esculentum	X	+	
Simmondsia chinensis	X		+
Sinapis alba	X		+
Endosperm			
Lactuca sativa	X		
Ricinus communis	X	+	+
Nicotiana tabacum			+
Perisperm			
Yucca brevifolia	X		+
Scutellum			
Zea mays	X	+	+

[a] Peroxisomes considered to be glyoxysomes because of tissue locale and morphology (electron microscopic) are scored X in column 1 (EM morph.). Those peroxisomes showing catalase reactivity with the 3,3′-diaminobenzidine (DAB) procedure are marked with a + in column 2 (DAB). Species from which glyoxysomes have been isolated on sucrose density gradients and shown to contain at least malate synthase and/or isocitrate lyase activity are marked in column 3 (Isolated) with a +, and those marked with a − were isolated but neither enzyme activity was assayed in the fractions.

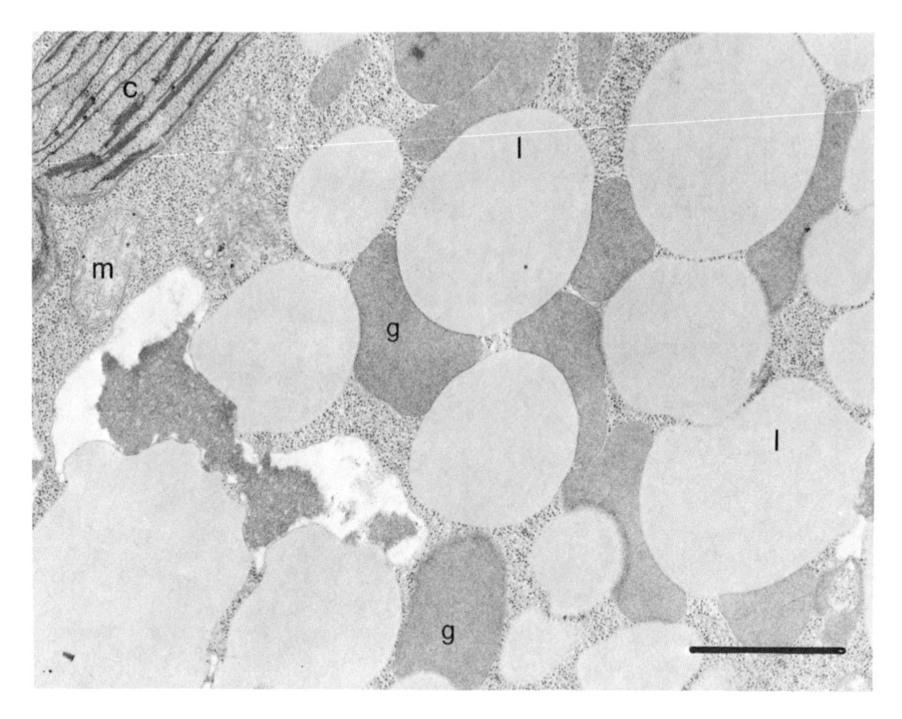

Fig. 2.13. Glyoxysomes (g) in intimate contact with lipid bodies (l) in a tomato cotyledon cell. Note the homogeneous nature of the glyoxysomal matrices and the variable shapes of the glyoxysomes. m, mitochondria; c, chloroplast. Bar = 1 μm. From P. J. Gruber (unpublished).

lustrating tomato glyoxysomes (Fig. 2.13). Glyoxysomes isolated on sucrose gradients appear ultrastructurally similar to those seen *in situ*, but the isolated particles tend to assume a more rounded shape (see Figs. 3.5, 3.6, and 3.7). The matrix portion of glyoxysomes shows some ultrastructural variability. In tomato cotyledons, a uniform nongranular matrix is often observed (Fig. 2.13), whereas the flocculent, more granular appearance of organelle matrices is common for glyoxysomes in other species (Figs. 2.14 and 2.16). Cytochemical analyses for malate synthase (Fig. 2.2), glycolate oxidase (Fig. 2.8), and catalase (Fig. 2.15) activities indicate that these enzymes are distributed in the matrix of the organelles. The type and occurrence of inclusions within glyoxysomes are variable among species. For example, crystalline arrays have not been observed in cotton, watermellon, or cucumber glyoxysomes, whereas they are often encountered in glyoxysomes of 3-day-old castor bean and sunflower (Fig. 2.14) storage tissues. An amorphous inclusion,

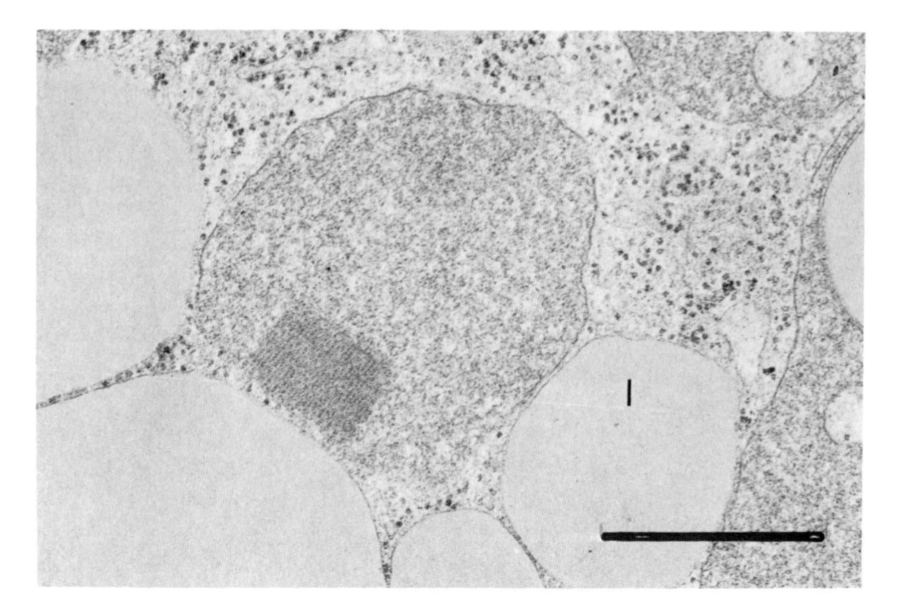

Fig. 2.14. Glyoxysome with a crystalline inclusion in a cell of a 4-day-old sunflower cotyledon. l, lipid body. Bar = 1 μm. From P. J. Gruber (unpublished).

similar to those seen in unspecialized and leaf peroxisomes, has been observed in glyoxysomes of essentially all species examined. There are insufficient ultrastructural data to make meaningful correlations between the type of glyoxysomal inclusion and their frequency or distribution among species.

Glyoxysomes in oil-storing cotyledons, pine megagametophyte, and *Dryopteris* fern spores exhibit views in thin sections that apparently are unique to glyoxysomes. They often are seen with portions of the cytoplasm apparently "trapped" within the organelles (Figs. 2.15 and 2.16). These "pockets" of cytoplasm, often containing recognizable ribosomes, actually are invaginations of cytoplasm into the glyoxysomes, continuous with the surrounding cytosol (Fig. 2.16, bottom). In some views, numerous invaginations extend into a single glyoxysome, illustrating their potentially high degree of pleiomorphy. As mentioned, this morphology appears to be unique to glyoxysomes, especially those in cotyledon cells which differentiate and become photosynthetic in response to light. In 1971, Trelease *et al.* quantitated the occurrence of these "pockets" in thin-section views of cucumber glyoxysomes (Table 2.4). The results clearly showed that the glyoxysomal invaginations occurred at a particular period of postgerminative growth, i.e., mostly

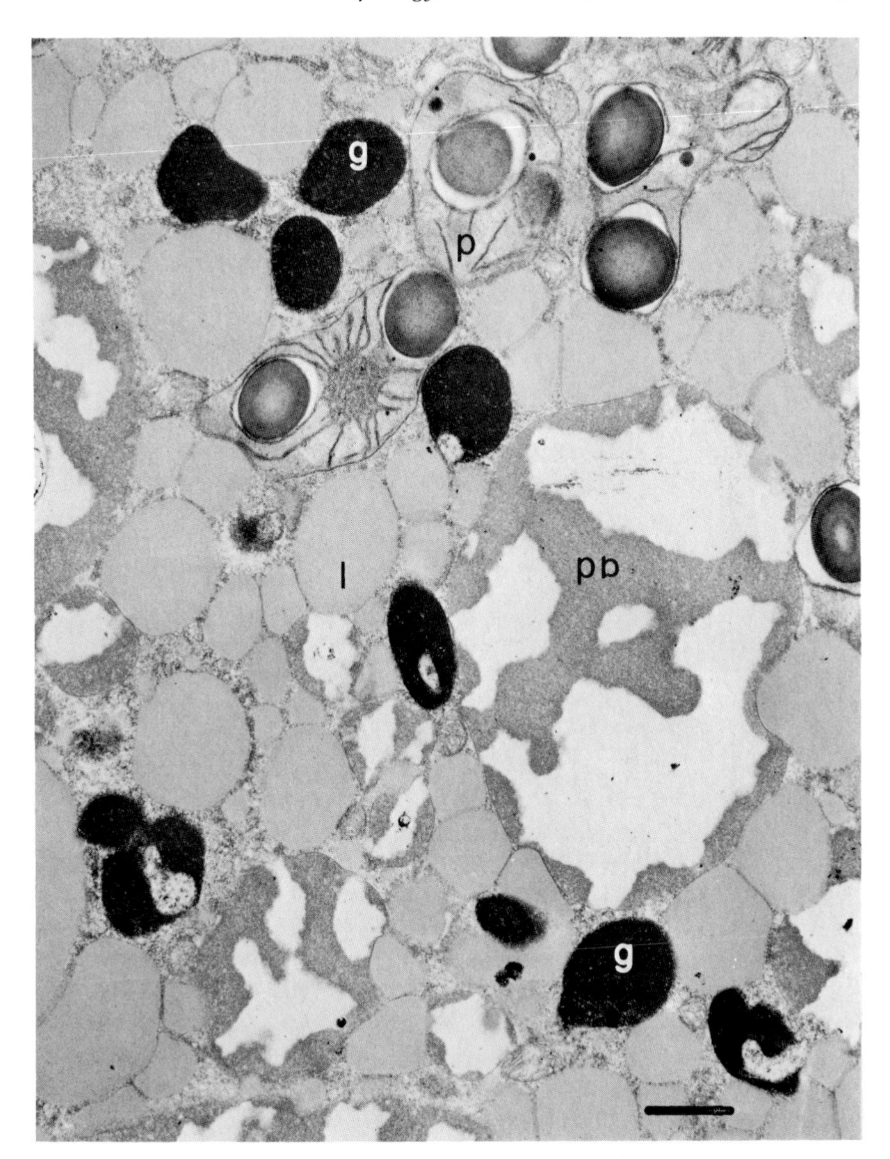

Fig. 2.15. Portion of a cucumber cotyledon cell from a 4-day-old seedling showing glyoxysomes (g) stained for catalase reactivity by the DAB procedure dispersed among the lipid bodies (l) and partially vacuolated protein bodies (pb). The reaction product is uniformly distributed in the glyoxysome matrix. Several of the glyoxysomes exhibit invaginations not containing catalase. Several plastids (p), among numerous lipid bodies (l), contain starch grains. Bar = 1.0 μm. From R. N. Trelease (unpublished).

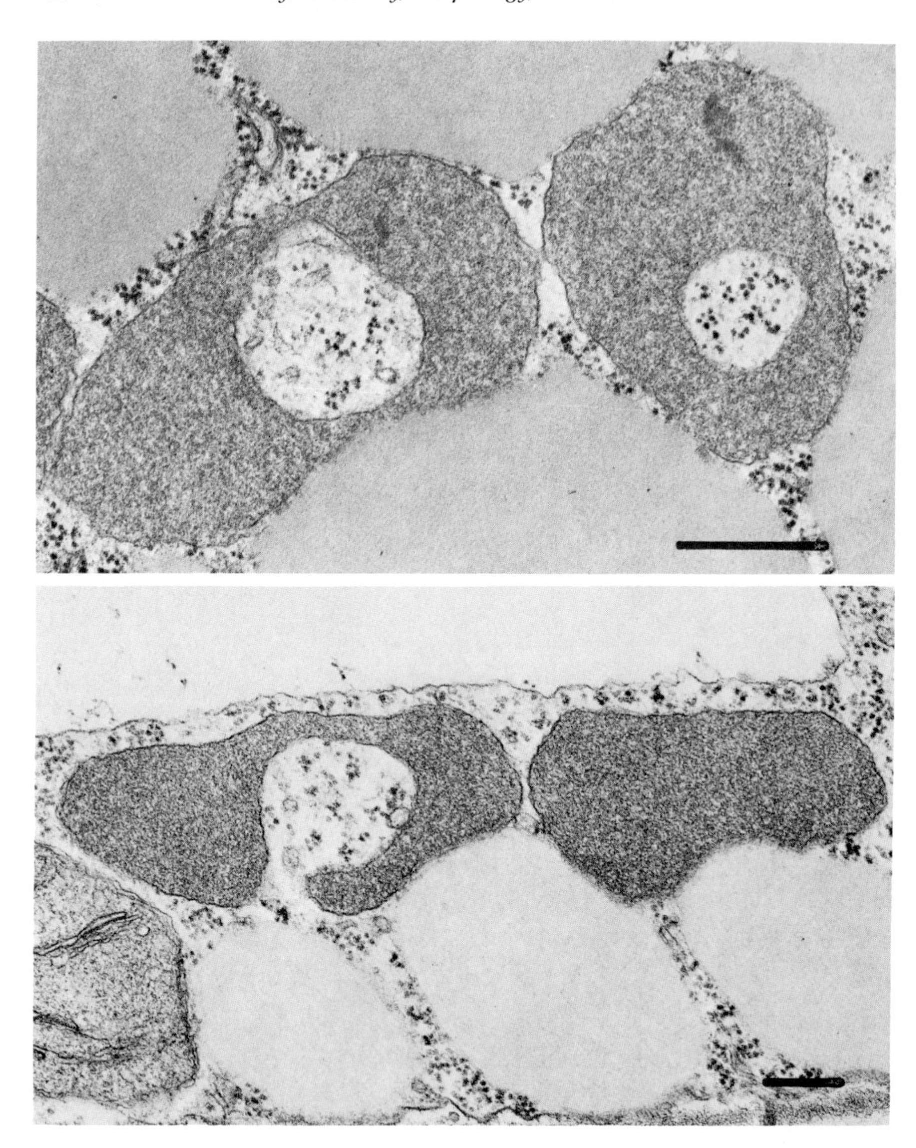

Fig. 2.16. Glyoxysomes in a palisade cell of a 4-day-old cucumber cotyledon. (Top, View of apparent pockets of "trapped" ribosomes within the glyoxysomes. (Bottom) View of similar glyoxysomes illustrating that the pockets actually are invaginations of a portion of the cytoplasmic groundplasm into the glyoxysome. An interesting and consistent observation is that the cytoplasmic material encompassed by the glyoxysomes is less electron opaque than the ground plasm outside the invaginated area (see also Fig. 2.15). Bars = 0.5 μm. From R. N. Trelease (unpublished).

TABLE 2.4

Percentage of Total Microbody Profiles Containing Invaginations when Viewed in Thin Sections at Different Stages of Postgerminative Growth in Cucumber Cotyledons (Seedlings Grown under a 12/12 hr Light–dark Cycle)[a]

Days	Percentage	Isocitrate lyase (nmol/min/cotyledon pair)	Glycolate oxidase (nmol/min/cotyledon pair)
2	0	50	0
3	10	82	1
4	68	80	3
5	11	10	8
6	4	4	9
7	0	0	9
10	0	0	10

[a] Adapted from Trelease *et al.* (1971).

near the period of peak glyoxylate cycle activity and during lipid degradation, with none observed after the depletion of lipids. Since this report in 1971, similar invaginations have been reported in glyoxysomes of numerous oil-storing cotyledons, but none has been described in unspecialized or leaf peroxisomes of higher plants, or in any algae or fungal peroxisomes. Such a definitive occurrence of a particular morphological feature of a specific type of organelle at a specific time period strongly suggests a structural–functional relationship. Accordingly, possible roles for these invaginations have been proposed, e.g., involvement in hydrolysis of lipid body triacylglycerols and/or a manifestation of the mechanism(s) involved in the glyoxysome-to-leaf peroxisome changeover that occurs in these cotyledons by some unresolved means (see Chapter 5). Castor bean glyoxysomes do not show these invaginations; thus, this morphological feature does not seem to be a general phenomenon among all glyoxysomes. Their function has not been resolved and hypotheses pertaining to their role are mostly speculative. However, their specific occurrence in glyoxysomes in cotyledons that can become photosynthetic, without any reported exceptions over the past 10 years, seems too coincidental for them not to reflect some functional or ontogenetic aspect of these glyoxysomes.

Glyoxysomes have been isolated and characterized from cotyledons of numerous species (Table 2.3), although the most extensive biochemical data have come from work on castor bean endosperm. A remarkable similarity in enzyme content and specific activity exists among glyoxysomes isolated from endosperm, megagametophyte, scutellum, and cotyledons (see Table 4.6). Comparative information is not available for these organelles in aleurone or perisperm tissues although isocitrate lyase was measured in glyoxysomes isolated from the perisperm of Joshua tree seeds. Peroxisomes have been isolated from barley and wheat aleurones, but there are no published data showing their possession of either malate synthase or isocitrate lyase, and in one instance the activities were reported as not present.

3. Photosynthetic Tissues

Green cotyledons and leaves assimilating carbon by the C_3, C_4, or CAM photosynthetic pathway have prominent peroxisomes in their mesophyll cells (Table 2.5). In C_4 plants, peroxisomes are more abundant on a per cell basis in bundle sheath cells than in mesophyll cells, but when one considers the number per tissue, the number of peroxisomes per mesophyll tissue approximates the number per bundle sheath tissue. It is likely that the enzyme complement of the peroxisomes in these tissues is not similar, i.e., bundle sheath peroxisomes probably contain active enzymes of the glycolate cycle, whereas the mesophyll peroxisomes may not. This is discussed in Chapter 4.

The prominence of peroxisomes in mature photosynthetic cells is evident from comparisons with the abundance of other organelles. For example, counts of organelles in thin sections of bean, oat, and wheat leaves indicate that peroxisomes are approximately one-half to one-third as numerous as mitochondria and chloroplasts, respectively (Frederick *et al.*, 1975). Leaf peroxisomes in C_3 plants were the largest microbodies reported in higher plant cells until the recent discovery of peroxisomes (up to 1.6 μm diameter) in soybean root cells adjacent to the infected nodule cells (see Fig. 4.9) (Newcomb and Tandon, 1981).

In both green cotyledons and leaves, the peroxisomes are closely appressed to chloroplasts and mitochondria (Figs. 1.7, 2.1, and 2.17 through 2.19), presumably reflecting their involvement with these organelles in photorespiration. Their shape is generally round to ovoid, but their contours often conform to the adjacent organelles (Figs. 2.17 through 2.19). DAB cytochemistry shows that catalase in peroxisomes lacking an inclusion is distributed throughout the matrix (Figs. 2.17, top, and 2.19). Crystal-containing peroxisomes show a deposition of DAB reaction product in the matrix and in the crystal (Fig. 2.1). Crystalline inclusions (Fig. 2.17, bottom) may be more common for leaf peroxisomes than other types of peroxisomes. The various noncrystalline inclusions

TABLE 2.5

Distribution of Peroxisomes in Green Cotyledons and Leaves of Angiosperms[a]

Tissues and species	EM morphology	DAB	Isolated
Cotyledons			
Citrullus vulgaris	X	+	+
Cucumis sativus	X	+	+
Cucurbita pepo (pumpkin)	X		+
Helianthus annuus	X	+	+
Lycopersicum esculentum	X	+	
Sinapsis alba	X	+	+
Spinacia oleacea			+
Leaves			
C_3			
Antirrhinum majus	X	+	
Atriplex patula			+
Avena sativa	X	+	
Beta vulgaris var. *cicla*			+
Beta vulgaris spp. *vulgaris*	X	+	+
Citrus aurantifolia	X		
Citrus jambhiri	X		
Dactylis glomerata	X	+	
Glycine max	X	+	
Hedychium coronarium	X	+	
Helianthus annuus			+
Hordeum vulgare	X	+	+
Lens culinaris			+
Linum usitatissimum	X		
Nicotiana sylvestris	X	+	
Nicotiana tabacum	X	+	+
Petunia sp.	X		
Phalaris arundinacea	X		
Phaseolus vulgaris	X	+	+
Phleum pratense	X		
Pisum sativum			+
Spinacia oleracea			+
Torenia fournieri	X		
Triticum aestivum	X	+	+
Triticum vulgare	X	+	+
Typha latifolia	X	+	
C_4			
Amaranthus hybridus			+
Amaranthus retroflexus	X		
Atriplex rosea			+
Chloris gayana	X		

(*continued*)

TABLE 2.5 (*Continued*)

Tissues and species	EM morphology	DAB	Isolated
Cynodon dactylon	X		
Cyperus rotundus	X		
Digitaria decumbens	X	+	
Digitaria sanguinalis	X		
Leptochloa dubia	X		
Saccharum CL 41-223			+
Saccharum officinarum	X	+	
Setaria viridis	X		
Sorghum sudanense	X	+	+
Zea mays	X	+	+
CAM			
Bryophyllum calycinum			+
Crassula lycopodioides			+
Crassula tetragona	X	+	
Echinocactus acanthodes (stem)	X		
Kalanchoë daigremontiana	X	+	
K. verticillata	X	+	
Sedum rubrotinctum	X	+	+
Unclassified (monocot sea grasses)			
Cymodocea rotundata	X		
C. serrulata	X		
Thalassia hemprichii	X		

[a] Peroxisomes identified in thin sections are scored X in column 1 (EM morph.). Those showing catalase reactivity with the 3,3′-diaminobenzidine (DAB) procedure are marked with a + in column 2 (DAB). Species from which peroxisomes have been isolated on sucrose density gradients and shown to contain at least catalase and/or glycolate oxidase activity are marked in column 3 (Isolated). Leaf species are grouped according to their type of photosynthetic pathway: C_3 pathway; C_4 pathway; CAM, crassulacean acid metabolism.

described previously in this chapter also occur in leaf peroxisomes, but it is difficult to estimate their comparative abundance. In festucoid grasses, narrow threads with a repeating unit seem to be unique. The peroxisome in an oat leaf cell illustrated in Fig. 2.18 has both the threadlike and crystalline inclusions within the same organelle. DAB cytochemistry also indicates that the threadlike inclusions sequester, or are composed of, some of the peroxisomal catalase (Frederick and Newcomb, 1971).

Complementing the electron microscopic observations of leaves and cotyledons from a wide range of species are numerous cell fractionation studies (Table 2.5). Isolated peroxisomes from these species all contain at least catalase and glycolate oxidase activity regardless of the main photosynthetic pathway (C_3, C_4, or CAM). It is interesting to note that no data, except cytochemical, are available on the enzyme content or biochemical role of the crystalline inclusions in leaf peroxisomes. Comparisons of enzyme constituents and roles in carbon metabolism are discussed in greater detail in Chapter 4.

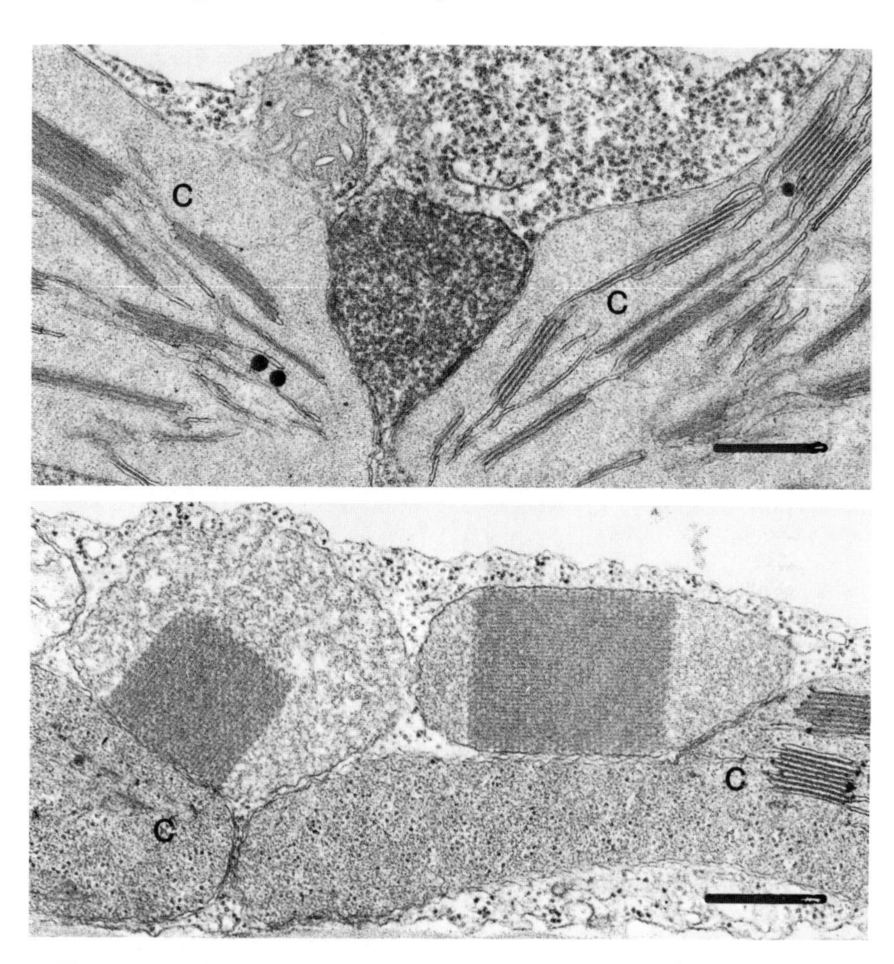

Fig. 2.17. Leaf-type peroxisomes in palisade cells of green cotyledons. (Top) Peroxisome in a 7-day-old cucumber cotyledon closely appressed to two chloroplasts (C). The glutaraldehyde-fixed tissue was treated with DAB and H_2O_2 for catalase reactivity. Storage lipid present earlier (e.g., Fig. 2.15) has been mobilized resulting in vacuolate cells with peroxisomes appressed to the chloroplasts. Bar = 0.5 μm. From R. N. Trelease (unpublished). (Bottom) Two peroxisomes in a 7-day-old sunflower cell appressed to chloroplasts (C). Crystalline inclusions are apparent in these peroxisomes. Bar = 0.5 μm. From P. J. Gruber (unpublished).

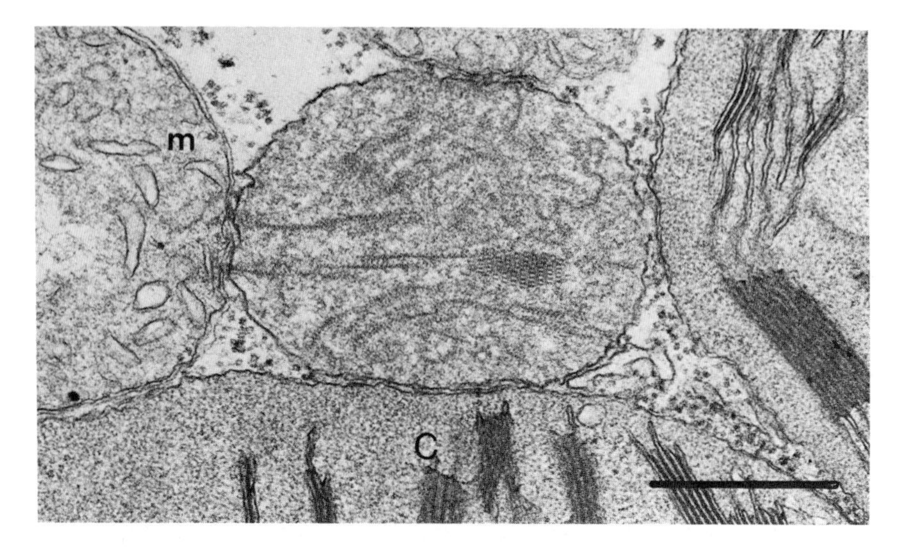

Fig. 2.18. Peroxisome in a light-grown oat leaf from a 9-day-old seedling. Both threadlike and crystalline inclusions occur in the same peroxisome. m, mitochondrion; c, chloroplast. Bar = 0.5 μm. From P. J. Gruber (unpublished).

B. Nonflowering Plants

1. Bryophytes and Nonflowering Tracheophytes

Microbodies have been observed with the electron microscope in green and nongreen cells of species representing essentially all the major groups of plants in this category (Table 2.6). Their ultrastructure and number per cell are variable when all the tissues are considered. Both crystalline and amorphous inclusions are observed in the matrices. Tissues from the various parts of mosses, ferns, and conifers have been examined most extensively. Except for the mosses and conifers, however, cytochemical tests for catalase reactivity have been few (Table 2.6). In all instances, the tests have been positive, suggesting that catalase is a constituent microbody enzyme. A comprehensive survey for glycolate oxidase activity in crude homogenates of these plants revealed oxidase activity in a tissue from every representative species examined (Frederick *et al.*, 1973). If one assumes that the oxidase is housed within the microbodies with the catalase (which is likely), then these groups of nonflowering plants would possess peroxisomes with the same basic enzyme content as found in essentially all tissues of flowering plants. Clearly, further work is needed to document this supposition, not only with those tissues in the species already examined, but in other tissues of the same species as well. For example, there is an obvious lack of information on the sporophyte generation (e.g., the fronds) in ferns (Table 2.6).

Biochemical information on peroxisomes isolated from these plants is available only for megagametophyte tissue of pine seeds (Ching, 1970, 1973; Lopes-Perez *et al.*, 1974). The research has shown that these organelles are similar in structure and function to glyoxysomes in oil-storage seed tissues of flowering plants. Malate synthase has been localized cytochemically in germinated spores of *Bryum* (Pais and Carrapico, 1979), but glyoxysomes have not been isolated from any of the moss spores. Isocitrate lyase and malate synthase activities have been measured in spore extracts from the ferns *Anemia phyllitidis, Pteris vittata, Dryopteris filix-mas,* and *Onoclea sensibilis,* but cytochemical localization of malate synthase has not been done and the activities have not been measured in isolated organelles. Gemmrich (1981) has done a thorough ultrastructural study of spores and protonema of *Anemia phyllitis* in which the association of microbodies with lipid bodies was correlated with the presence of glyoxysomal enzyme activities. Thus, there is some evidence that glyoxysomes occur in these nonflowering plants.

Evidence for "leaf-type" peroxisomes comes only from ultrastructural

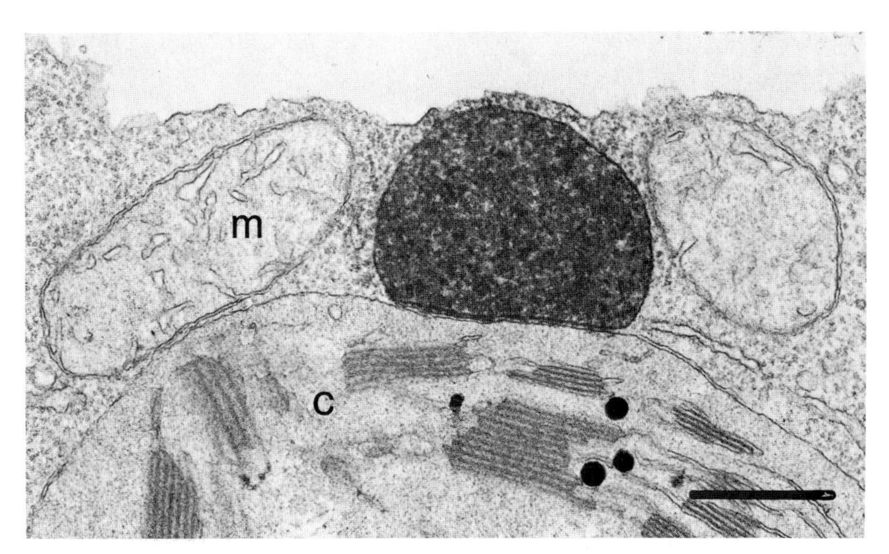

Fig. 2.19. Peroxisome in a tobacco leaf mesophyll cell adjacent to a chloroplast (c). The tissue segment was incubated in the DAB medium; reaction product is confined to the microbody and is not observable in mitochondrion (m) or chloroplast (c). Bar = 0.5 μm. [From Newcomb and Frederick (1971). *In* Photosynthesis and Photorespiration. (M. D. Hatch, C. B. Osmond, and R. O. Slayter, eds). Reprinted by permission of John Wiley & Sons, Inc.]

TABLE 2.6

Species of Bryophytes and Nonflowering Tracheophytes Shown to Possess Peroxisomes by Electron Microscopic Examination of Thin Sections[a]

Species	DAB
Bryophyta	
Mosses	
Aulacomnium sp. (gemmae)	
Bryum caespiticium (capsule)	+
Bryum capillare (germinated spores)[b]	+
Fissidens limbatus (capsule)	+
Funaria hygrometrica (capsule)	+
F. hygrometrica (gametophyte)	+
F. hygrometrica (antheridium)	
Polytrichum commune (gametophyte)	+
Liverworts	
Marchantia polymorpha (thallus)	+
Hornworts	
Anthoceros sp. (sporophyte)	
Tracheophyta	
Whisk-ferns	
Psilotum nudum (sporophyte)	
Horsetails	
Equisetum arvense (sporophyte)	+
Ferns	
Anemia phyllitidis (spores and gametophyte)	+
Dryopteris filix-mas (spores and gametophyte)	
Osmunda claytoniana (antheridium)	

(*continued*)

studies. Microbodies (DAB stained) were observed closely associated with chloroplasts in pine needles (Chabot and Chabot, 1974), *Polytrichum commune* leaves (Hebant and Marty, 1972), and *Anemia protonema* (Gemmrich, 1981). Cytochemical evidence for glycolate oxidase in peroxisomes adjacent to chloroplasts is available for germinated, green *Bryum* spores (Pais and Carrapico, 1979). Taken together, it may be assumed that peroxisomes possessing photorespiratory enzyme activities are present in green tissues of these plants, but the lack of any cell fractionation experiments makes this a tenuous conclusion.

In summary, catalase-containing peroxisomes probably have a widespread occurrence in various tissues of bryophytes and nonflowering tracheophytes, but biochemical data are lacking on their metabolic function(s). Pine seeds have well-defined glyoxysomes, but the occurrence of glyoxysomes within other plants of this category is unknown. Evi-

TABLE 2.6 (*Continued*)

Species	DAB
Polypodium vulgare (spores and gametophyte)	
Pteridium aquilinum (gametophyte)	
Thelypteris normalis (antheridium)	
Conifers	
Abies balsamea (needles)	
Juniperus osteosperma (needles)	
Picea abies (needles)	+
Pinus jeffreyi (megagametophyte)[c]	
Pinus mariana (needles)	+
Pinus pinea (megagametophyte)[c]	
Pinus ponderosa (megagametophyte)[c]	+
Pinus strobus (needles)	+
Pinus sylvestris (megagametophyte)	
Cycads	
Dioon edule (leaf)	+
Zamia latifoliata (leaf)	+

[a] Peroxisomes demonstrated to have cytochemical reactivity for catalase by the 3,3′-diaminobenzidine (DAB) procedure are listed with the + symbol.

[b] Peroxisomes in spores cytochemically stained *in situ* for malate synthase reactivity at an early stage of germination and for glycolate oxidase reactivity at a later stage (Pais and Carrapico, 1979).

[c] Peroxisomes from megagametophyte tissues isolated and well-characterized by enzyme analyses as glyoxysomes similar to those found in oilseeds of flowering plants (Ching, 1970, 1973; Lopes-Perez *et al.*, 1974).

dence for the existence of leaf-type peroxisomes is indicated only by a few ultrastructural microbody–chloroplast associations.

2. Fungi

Electron microscopic observations of essentially all groups of fungi at various developmental stages indicate that peroxisomes are ubiquitous organelles in fungal cells (Tables 2.7 and 2.8). Morphological identification of peroxisomes in fungal cells have been difficult due to the common occurrence of similar, single-membrane-bounded, but not peroxisomal organelles. Furthermore, application of the cytochemical DAB procedure for catalase localization often has been unsuccessful, especially with filamentous fungi. The standard procedure (described earlier in this chapter) has yielded negative results in several organisms (see Table 2.7 for examples). However, more recent modifications of the standard procedure have resulted in positive staining in species where

TABLE 2.7

Distribution of Peroxisomes in Fungi Revealed by Electron Microscopic Examination of Thin Sections[a]

Species[b]	DAB reaction
Chytridiomycetes	
Blastocladiella emersonii	+
Entophlyctis sp.	+
Entophlyctis variabilis	+
Oomycetes	
Achyla ambisexualis	+
Aphanomyces euteiches	
Phytophthora palmivora	+
Phythium ultimatum	−
Zygomycetes	
Rhizopus arrhizus	
Rhizopus nigricans	−
Ascomycetes	
Aspergillus tamarii	
Aspergillus niger	+
Botrytis cinerea	−
Candida albicans (*n*-alkane)	+
C. guilliermondii (*n*-alkane)	+
C. intermedia (*n*-alkane)	+
C. lipolytica (*n*-alkane)	+
C. tropicalis (*n*-alkane)	+
Cladosporium cucumerinum	
Cladosporium resinae	
Hansenula polymorpha (methanol)	+
Helminthosporium turcicum	
Kloeckera sp. (methanol)	+
Melampsora lini	−
Neurospora crassa	−
Pichia pinus (methanol)	+
Saccharomyces cerevisiae (glycerol)	+
S. cerevisiae (glucose, galactose)	−
Stemphylium sarcinaeforme	
Whetzelinia sclerotiorum	−
Deuteromycetes	
Fusarium oxysporum	−
Sclerotium rolfsii	−
Basidiomycetes	
Phanerochaete sp.	
Poria contigua	+
Puccinia helianthus	−
Rhizoctonia solani	
Uromyces phaseoli	−

[a] Fungal species shown to have positively stained peroxisomes with 3,3′-diaminobenzidine (DAB) are indicated with a +; − signifies negative reactivity.

[b] Taxonomic system according to Alexopoulos (1962).

TABLE 2.8

Occurrence of Peroxisomes in Plant Parasitic or Pathogenic Fungi and the Association of Peroxisomes with Different Developmental Stages of the Fungus[a]

Fungus	Hyphae	Zoospore or cyst[b]	Asexual spore or structure[b]	Sexual spore or structure	Parasitic or pathogenic stage
Myxomycota					
Plasmodiophoromycetes					
Plasmodiophora brassicae		+[c]			
Eumycota					
Mastigomycotina					
Chytridiomycetes					
Rozella allomycis		+			+
Oomycetes					
Aphanomyces euteiches	+	+			+
Bremia lactucae			+		
Phytophthora capsici		+			
Phytophthora cinnamomi			+		
Phytophthora megasperma		+			++
Phytophthora parasitica		+			
Phytophthora palmivora		++			
Phytophthora infestans		++			+
Pythium spp.	+++				
Zygomycotina					
Zygomycetes					
Endogone sp.			++		
Rhizopus arrhizus	+++				
Ascomycotina					
Hemiascomycetes					
Taphrina maculans			++		
Loculoascomycetes					
Cymadothea trifolii	+				+
Helminthosporium turcicum	+				
Spilocaea pomi					+
Stemphylium sarcinaeforme			++		
Cochliobolus carbonus			++		+
Drechslera sorokiniana			+++		
Plectomycetes					
Ceratocystis ulmi	+				
Aspergillus tamarii	+++				
Pyrenomycetes					
Fusarium oxysporum f. sp.	+++				

(*continued*)

TABLE 2.8 (*Continued*)

Fungus	Hyphae	Zoospore or cyst[b]	Asexual spore or structure[b]	Sexual spore or structure	Parasitic or pathogenic stage
conglutinans					
Fusarium oxysporum f. sp. *lycopersici*	+				+++
Discomycetes					
Botrytis cinerea	++				
Whetzelinia sclerotiorum	++				+
Basidiomycotina					
Teliomycetes					
Hemilicia vastatrix	+				+
Melampsora lini	+				
Puccinia graminis f. sp. *tritici*			+		
Puccinia helianthi					+
Tilletia caries				+++	
Uromyces phaseoli var. *typica*			++		
Hymenomycetes					
Armillaria mellea	+				
Rhizoctonia solani	+				
Sclerotium rolfsii	++				+++

(*continued*)

the cytochemistry previously was negative (Maxwell *et al.*, 1977). Since catalase is associated with all peroxisomes that have been isolated from fungi (Table 2.9), and the DAB procedure is positive for peroxisomes in diverse groups (Table 2.7), it is probable that catalase is a constituent enzyme of fungal microbodies. As a consequence of the earlier difficulties in morphologically identifying peroxisomes in fungi, the peroxisomes assumed a host of different names, e.g., cytosome, Stüben bodies, lysosome or spherosome (as were higher plant peroxisomes), side-body matrix, Woronin bodies, etc. Considerable ultrastructural and cytochemical data are now available, however, to verify the widespread occurrence of peroxisomes in the various fungal cells.

The occurrence of a H_2O_2-generating oxidase within peroxisomes of

TABLE 2.8 (*Continued*)

Fungus	Hyphae	Zoospore or cyst[b]	Asexual spore or structure[b]	Sexual spore or structure	Parasitic or pathogenic stage
Deuteromycotina					
Hyphomycetes					
Cercosporella herpotrichoides	+				+
Cladosporium cucumerinum	++		+++		+++
Cladosporium fulvum	+++				+++[d]
Coelomycetes					
Botryodiplodia theobromae			+++		
Colletotrichum coccodes	++		++		
Colletotrichum graminicola			++		++
Colletotrichum lindemuthianum	+				+
Pyrenocheta terrestris	++				++

[a] From Maxwell *et al.* (1977). [Reproduced with permission, from *Ann. Rev. Phytopathol.* (1977) **15,** 119–134.]

[b] Includes germinating structures.

[c] Relative numbers of microbodies: +, few; ++, a moderate number; +++, many.

[d] Hyphae in the resistant tomato cultivar have more peroxisomes than hyphae in the susceptible tomato cultivar.

the various fungi is not as well documented. The list in Table 2.9 shows that assays for any oxidase activity were not done in approximately half of the cell fractionation studies. Moreover, data are lacking on oxidase activity in homogenates or in sediments from differential centrifugation of fungal homogenates. However, van Dijken and Veenhuis (1980) and Bringer *et al.* (1980) recently provided the first cytochemical evidences for oxidases (glucose and methylamine, and methanol oxidase, respectively) in peroxisomes of filamentous fungi (*Aspergillus niger* and *Poria contigua*, respectively). It is generally assumed that catalase exists within these organelles to breakdown H_2O_2 generated by oxygen-coupled oxidative enzymes. Accepting this assumption, and considering recent studies that negative results for oxidase activity have not been reported

TABLE 2.9

Fungal Species from Which Peroxisomes Have Been Isolated by Density-Gradient Centrifugation[a]

Species	Enzyme activity[b]			
	α-HAO	AO	DAO	CAT
Chytridiomycetes				
Blastocladiella emersonii sporangia				+
Botryodiplodia theobromae spores				+
Entophlyctis sp. zoospores				+
Oömycetes				
Phytophthora palmivora				+
Ascomycetes				
Aspergillus tamarii (ethanol)				+
Candida boidinii (methanol)		+		+
Candida tropicalis pk233 (n-alkane)			+	+
Candida utilis (amines)				+
Hansenula polymorpha (methanol)	+	+	+	+
Hansenula polymorpha (amines)				+
Kloeckera sp. (methanol)		+	+	+
Neurospora crassa (acetate)				+
Saccharomyces cerevisiae (ethanol & glucose)	+			+
Saccharomyces cerevisiae (lactose)	+		+	+
Deuteromycetes				
Sclerotium rolfsii (pathogenic)				+
Basidiomycetes				
Coprinus lagopus (acetate)				

[a] Enzyme activity associated with the peroxisomes is indicated by a "+". The carbon or nitrogen source on which the fungi were grown is indicated where appropriate. (see Chapter 4 for other details)

[b] Enzyme abbreviations: α-HAO, α-Hydroxyacid oxidase; AO, Alcohol oxidase; DAO, D-amino acid oxidase; CAT, Catalase.

in fungi, it seems likely that catalase and an oxidase are basic constituent enzymes of peroxisomes in fungi as they are in higher plants. From what is known already about the involvement of peroxisomes in varied metabolic roles (Chapter 4), it appears that fungal peroxisomes assume more varied roles than those in higher plants.

Fungal peroxisomes assume shapes as variable as their number per cell (Table 2.10), the latter being influenced both by the age of cells and the substrate or carbon source upon which they are grown (Maxwell *et al.*, 1975). Examples of peroxisomes in a zoospore and hypha of *Allomyces* are illustrated in Fig. 2.20. Zoospore peroxisomes are nearly spherical with an irregular outline. In the hyphal cell, a small pleiomor-

TABLE 2.10

Number of Organelles in Hyphal Tip Cells of Fungi Grown on Medium-coated Glass Microscope Slides[a]

Fungus	Carbon source	Average no. organelles/7.5 μm length of hypha					Ratio of microbodies to mitochondria	No. of 7.5 μm lengths observed
		Microbodies	Mitochondria	Vacuoles	Nuclei	Lipid bodies		
Neurospora crassa	Ethanol	3.1	14.6	12.8	1.8	—[b]	0.21:1	77
N. crassa	Glucose	2.1	16.7	20.4	2.4	—[b]	0.13:1	17
Pythium ultimum	Polypectate	2.1	8.0	10–25[d]	0.7	0.3	0.26:1	309
P. ultimum	cmc[c]	1.2	3.9	10–25[d]	0.5	0.1	0.31:1	134
P. ultimum	Ethanol	0.1	2.6	10–25[d]	0.5	0.9	0.04:1	107
Rhizopus arrhizus	Ethanol	3.1	3.3	13.0	0.6	—[b]	1:1	51
Sclerotium rolfsii	Polypectate	1.0	3.6	1.7	0.5	—[b]	0.28:1	164
S. rolfsii	Glucose	0.2	5.4	4.1	0.4	—[b]	0.037:1	49
Whetzelinia sclerotiorum	cmc[c]	2.0	15.2	6.8	0.8	1.0	0.13:1	26

[a] Counts were made from a single longitudinal thin section of each hyphal tip and several tips of each fungus were observed. From Maxwell *et al.* (1975).

[b] Lipid bodies were less than 1 μm diameter and were not counted.

[c] Carboxymethyl cellulose.

[d] Because of the large numbers of vacuoles in the hyphae of *P. ultimum*, numbers are only estimates.

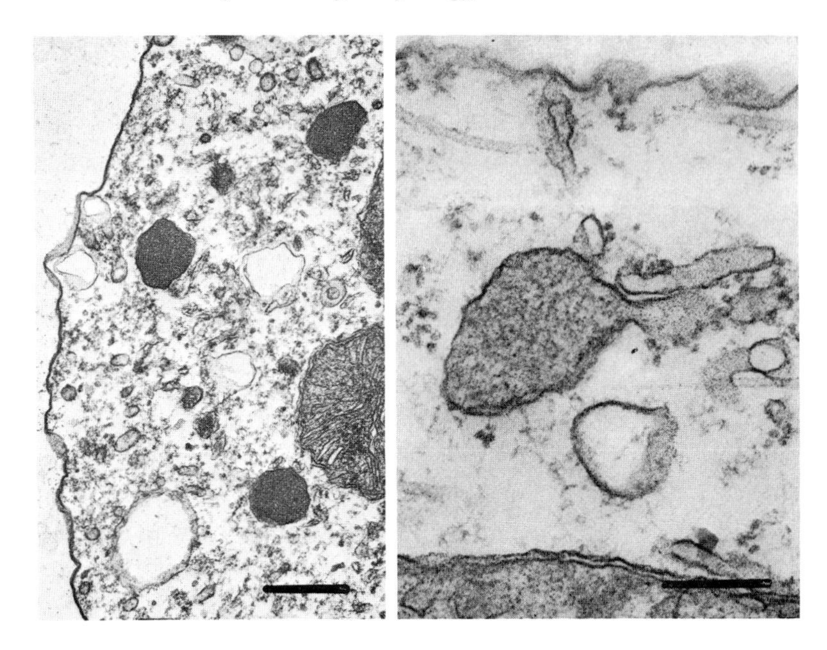

Fig. 2.20. Unspecialized peroxisomes in fungi. (Left) Three irregular-shaped peroxisomes in the peripheral cytoplasm of an *Allomyces macrogynus* zoospore. Bar = 0.5 μm. From V. Sandstedt (unpublished). (Right) Peroxisome in the hyphal cytoplasm of *Allomyces macrogynus* showing an apparent attachment and association with RER profiles. Bar = 0.25 μm. From M. J. Powell (unpublished).

phic peroxisome is shown associated with, and possibly attached to, segments of rough endoplasmic reticulum. This type of relationship often is observed in fungal cells. A characteristic association of peroxisomes with lipid bodies is shown in Fig. 2.21 (left), where the combination forms a so-called microbody–lipid globule complex (MLC). Based on the organization of the MLC in chytridiomycetous zoospores, Powell (1978) suggests that it may provide an additional criterion by which the phylogeny of aquatic fungi can be evaluated. Elongated peroxisomes are also common in fungi (Fig. 2.21, right). As an extreme in peroxisome morphology, an elaborate peroxisome is depicted in Fig. 2.22. It is apparently a large sheet-like structure, observed in one thin section as an elongate organelle surrounding the large vacuole within the sporangium (Fig. 2.22). In methanol-grown yeast cells and in mycelia of *Aspergillus niger* grown in fermenter cultures, the peroxisomes proliferate and enlarge to about 1.7 μm in diameter. This is larger than the leaf peroxisomes in C_3 plants and soybean nodule peroxisomes which are about 1.5 μm in diameter. Different types of inclusions also are seen within fungal peroxisomes, but less frequently than observed in flowering plants. These include both crystalline and noncrystalline types. Crystalline inclusions are especially well developed in methanol-grown yeast cells and in *Cladosporium resinae* hyphae grown on glucose.

Comparing the list in Table 2.9 with those in Tables 2.7 and 2.8 reveals that considerably more data on fungal peroxisomes have come from electron microscopy than from cell fractionation studies. A main reason for this disparity is the common difficulty encountered when attempting to isolate organelles from fungal cells. The majority of the information on isolated particles has come from the work on yeast cells cultured on either methanol or *n*-alkane growth media. The data on these organisms are considered in detail in Chapter 4 as is the information on the occurrence and enzyme content of glyoxysome-like peroxisomes in fungi.

In summary, it is clear that microbodies are common organelles in all fungi, and that catalase is probably a constituent enzyme of these organelles. Although only limited data are available on the widespread distribution of H_2O_2-producing oxidases, the available information indicates the enzymes are associated with the peroxisomes. More work is needed to identify specific peroxisomal functions in the various groups of fungi. Perhaps the greatest obstacle will be overcoming the problems

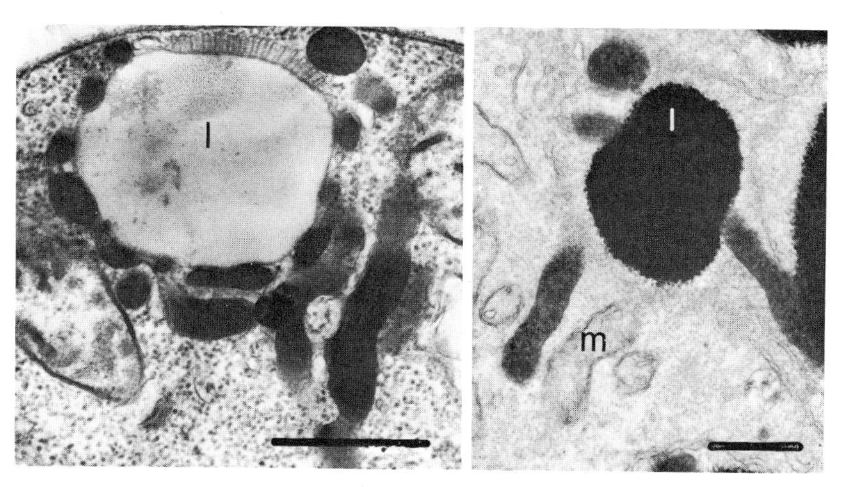

Fig. 2.21. Peroxisomes (glyoxysomes?) stained for catalase reactivity by the DAB procedure in fungi. (Left) Peroxisomes surrounding a lipid body (l), forming a "microbody-lipid globule complex" (MLC) in a recently encysted zoospore of *Rhizophydium spherotheca*. Bar = 0.5 μm. (Right) Elongate peroxisomes near a darkly stained lipid body (l) in a prosporangium of *Polyphagus englenae*. The lipid body appears electron opaque due to *en bloc* uranyl acetate staining. m, mitochondria. Bar = 0.5μm. From M. J. Powell (unpublished).

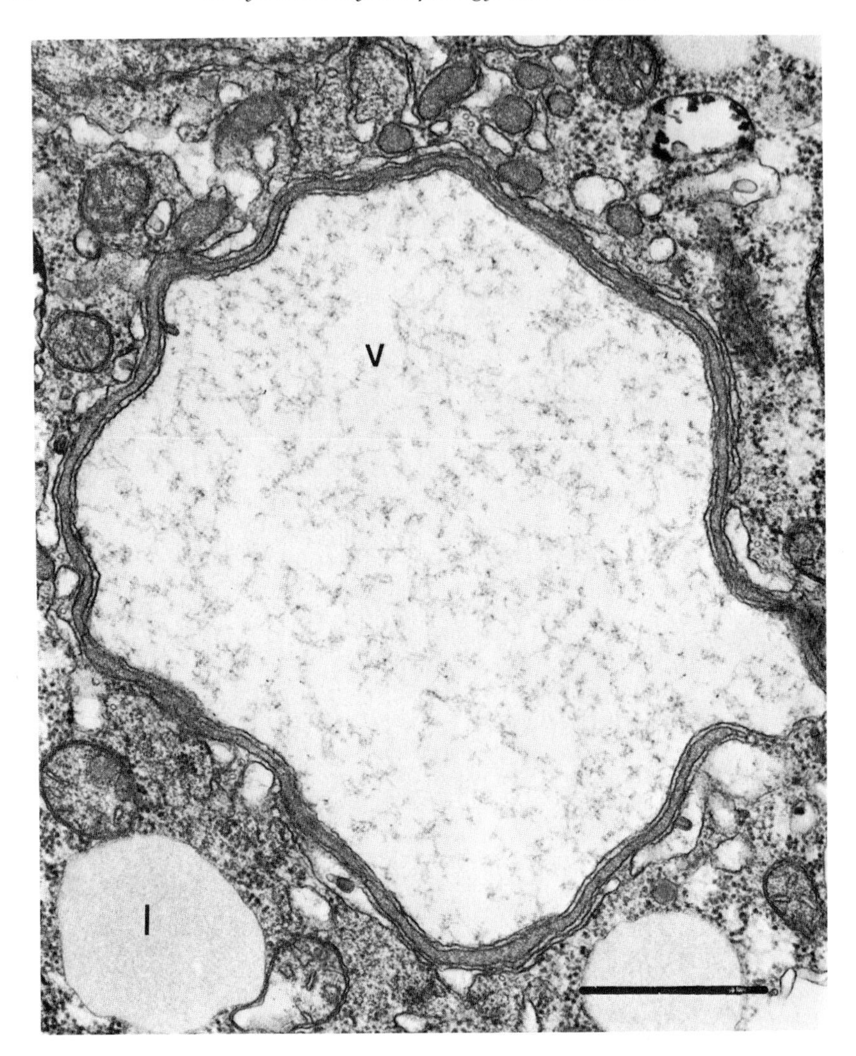

Fig. 2.22. An elaborate peroxisome in a sporangium of *Chytriomyces hyalinus*. In this view, an elongate peroxisome surrounds a large central vacuole (v). m, mitochondrion; n, nucleus; l, lipid body. Bar = 1 μm. From M. J. Powell (unpublished).

inherent in isolating fungal organelles. In those instances where peroxisomal function is well documented, the details are discussed relative to their type of metabolism (Chapter 4).

3. Algae

Peroxisomes have been observed with the electron microscope in at least one species of every taxonomic division of algae (Table 2.11) except

TABLE 2.11

Algal Species Shown to Possess Peroxisomes by Electron Microscopic Examination of Thin Sections[a]

Species[b]	DAB	Species	DAB
Chlorophyta		*Schizomeris* sp.	
Acetabularia mediterranea	+	*Spirogyra* sp.	+
Brachteococcus cinnabarinus	+	*Spongiochloris excentrica*	
Chlamydomonas dysosmos	+	*Stichococcus* sp.	−
C. reinhardtii	−	*Stigeoclonium* sp.	+
C. segnis		*Trichosarcina* sp.	
Chlorella pyrenoidosa		*Ulothrix fimbriata*	
C. sorokiniana (wild type)		*Ulva* sp.	
C. vulgaris (wild and mutant types)	+	*Uronema* sp.	
		Zygnema sp.	
Chlorochytrium sp.		Charophyta	
Chlorogonium elongatum	−	*Chara* sp.	
Cladophora sp.		*Nitella flexilis*	+
Closterium littorale		*Tolypella* sp.	
Coleochaete nitellarum		Euglenophyta	
C. scutata	+	*Euglena gracilis*	−
Cosmarium botrytis	+	Phaeophyta	
Dictyosphaerium pulchellum		*Ectocarpus* sp.	+
Draparnaldia sp.		*Fucus* sp.	
Dunaliella tertiolecta		*Giffordia* sp.	
Enteromorpha sp.		*Laminaria* sp.	
Eremosphaera viridis	+	Chrysophyta	
Gonium pectorale		*Botrydium granulatum*	
Hormidium sterile		*Chrysocapsa epiphytica*	+
Hyalotheca sp.		*Gonyostomum semen*	
Klebsormidium flaccidum	+	*Ochromonas malhamensis*	
Mesotaenium caldariorum		*Vacuolaria virescens*	
Micrasterias denticulata		*Vaucheria* sp.	
M. fimbriata	+	Pyrrhophyta	
Microspora sp.		*Amphidinium* sp.	
Netrium digitus	+	*Aureodinium* sp.	
Netrium sp.		*Cachonia niei*	−
Oedogonium cardiacum		*Ceratium hirundinella*	−
Platymonas subcordiformis		*Glenodinium* sp.	
Polytoma uvella		*Gymnodinium* sp.	
Polytomella caeca	−	*Heterocapsa* sp.	
Protosiphon sp.	+	*Katodinium* sp.	
Prototheca stagnora		*Oxyrrhis marina*	−
P. zopfii		*Peridinium* sp.	
Radiofilum sp.		*Prorocentrum* sp.	
Scenedesmus obliquus		*Scrippsiella sweeneyae*	−
S. obtusiusculus		*Woloszynskia micra*	−

(*continued*)

TABLE 2.11 (*Continued*)

Species[b]	DAB	Species	DAB
		Rhodophyta	
		Bangia fuscopurpurea	+
		Palmeria palmata	+
		Porphyridium cruentum	+
		P. purpureum	+
		Unknown taxonomic origin	
		Cyanophora paradoxa	

[a] Peroxisomes with catalase activity demonstrated by 3,3′-diaminobenzidine (DAB) cytochemistry are designated with a +; the − symbol indicates a negative reaction; those species listed without any symbol have not been tested for catalase reactivity.

[b] Taxonomic system modified from Bold and Wynne (1978).

Cyanophyta (blue-green algae). The greatest number of observations have been in the Chlorophyta, largely due to a comprehensive survey of the organelles in green algae conducted by Silverberg (1975). Much variation in the size, shape, and number of peroxisomes per cell is seen in different algae. Their size ranges from approximately 0.1 to 1.8 μm in the greatest diameter, although the range most commonly observed is from 0.2 to 0.5 μm (Table 2.12). Peroxisomes in *Nitella flexillus* are the largest in diameter of the algal peroxisomes examined (0.4–1.8 μm; Silverberg, 1975) and appear to be the largest reported in the plant or animal kingdom. Elongate peroxisomes may reach nearly 2.0 μm. In general, the peroxisomes are much smaller and less numerous than mitochondria (Table 2.12). In some instances, only one peroxisome per cell has been seen, e.g., in *Hormidium sterile* and *Klebsormidium flaccidum* (Fig. 2.24), and these single peroxisomes vary in size. The majority of peroxisomes studied do not exhibit distinctive inclusions such as those frequently seen in flowering plants and certain fungi. In many algae, the matrix material has an electron density greater than the surrounding cytosol. Inconspicuous organelles with low electron opacity such as the unspecialized peroxisomes occurring in higher plant parenchyma cells also are common in certain algal species. The latter ultrastructural characteristic often contributed to peroxisomes being overlooked in both the higher plant and algal cells, and in the case of certain algae led researchers in the early 1970s to conclude that microbodies (peroxisomes) did not exist in these algae. Association of peroxisomes with profiles of RER is frequently observed, and close spatial associations with lipid bodies, chloroplasts (Fig. 2.23), and/or mitochondria (Fig. 2.24) also have been seen in a number of algal species. An apparently unique locale for algal peroxisomes is shown in *Klebsormidium* (Fig. 2.24, left) where a single organelle is positioned between the nucleus and chloroplast.

Algal peroxisomes are more widely distributed than previously thought, and now their functional role is of interest. The list in Table 2.11 shows 19 species positive and 10 species negative for DAB cytochemical reactivity in thin sections. The negative results have been attributed to several factors. (1) Some autotrophic algal species possess a glycolate dehydrogenase enzyme rather than the glycolate oxidase, and since the dehydrogenase does not couple directly to oxygen to produce H_2O_2, catalase would not be required. (2) The standard procedure used for cytochemically localizing catalase in flowering plants is not suitable for certain algal species (a similar situation encountered with fungal species). (3) The presence of peroxisomes and/or the compartmentation of catalase within these organelles varies depending on the substrate induction, growth conditions, and/or stage of development of the organism. (4) The catalase activity is too low to detect in small cells with few peroxisomes. The first factor can be negated, since biochemical ac-

TABLE 2.12

Size, Shape, and Relative Frequency of Organelle Profiles in Eight Chlorophycean Representatives[a]

Species	Size (μm) (based on 100 microbodies)	Shape	Number of profiles per cell section	
			Mitochondria	Microbodies
Chlamydomonas dysosmos	0.25–0.90	Spherical, ovate	4.30 ± 0.41	0.42 ± 0.08
Dictyosphaerium pulchellum	0.15–0.30	Ovate, pyriform	1.17 ± 0.13	0.20 ± 0.07
Protosiphon sp.	0.18–0.22 (diameter) 0.40–0.80 (length)	Ellipsoid, dumbbell	7.73 ± 1.2	0.80 ± 0.21
Scenedesmus obliquis	0.15–0.25	Ovate	4.13 ± 0.39	0.67 ± 0.09
Hormidium sterile	0.20–0.45 (diameter) 0.90–1.2 (length)	Ovate, dumbbell	1.79 ± 0.24	0.79 ± 0.11
Stichococcus	0.18–0.40	Ovate, ellipsoid	1.40 ± 0.18	0.20 ± 0.07
Cosmarium botrytis	0.32–0.53	Spherical, ovate	1.73 ± 0.24	0.17 ± 0.07
Netrium digitus	0.35–0.55	Spherical, ovate	2.57 ± 0.26	0.07 ± 0.05

[a] From Silverberg (1975).

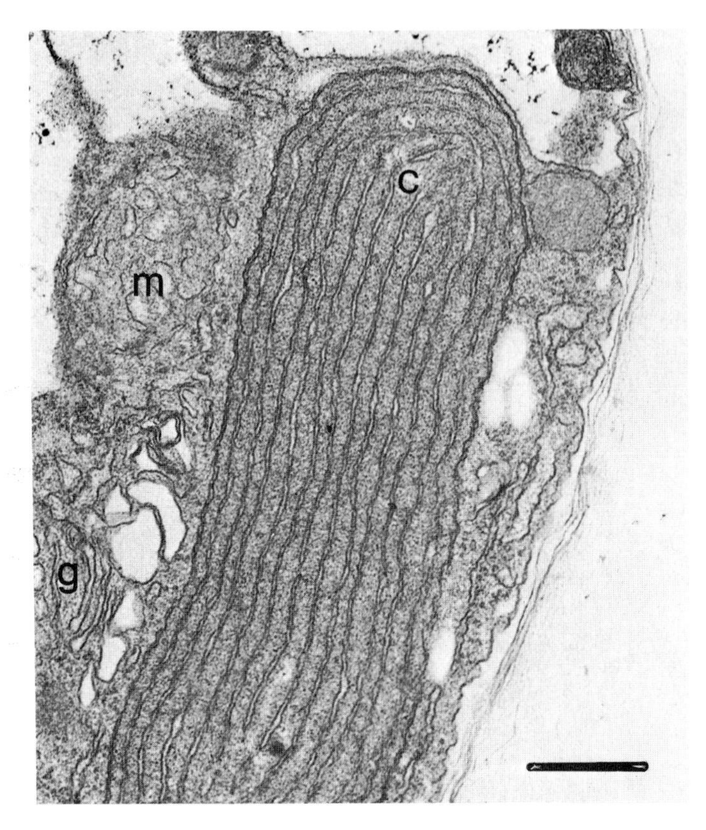

Fig. 2.23. Peroxisome adjacent to the chloroplast (c) in a germinating carpospore of the red alga, *Bangia fuscopurpurea*. Several peroxisomes occur in these cells, each having a similar spherical shape and size, adjacent to a large chloroplast. g, Golgi body; m, mitochondrion. Bar = 0.5 μm. From H. P. Lin (unpublished).

tivity of catalase in homogenates and cytochemical reactivity in thin sections have been found in algal species regardless of the type of glycolate-metabolizing enzyme present (Frederick *et al.*, 1973; Silverberg, 1975). The second factor has been substantiated; in many instances where the standard cytochemical procedure has failed to localize activity, modification of the procedure has yielded stained particles (Silverberg, 1975). Factor (3) is becoming more apparent as more work is being done, and certainly factor (4) remains a possibility, although unlikely since catalase activity is easily measured in higher plant tissues with few peroxisomes. Thus, the negative results for DAB cytochemistry can be attributed to factors (2) and (3). Previously thinking as outlined in (1) certainly was reasonable in the absence of definitive data, which are now available. On the basis of the mounting positive evidence, and technical explanation for most of the negative results, it seems that catalase is a constituent enzyme of algal peroxisomes with possibly few exceptions.

Table 2.13 lists the algal species from which microbodies have been isolated on sucrose density gradients. The same situation that exists for the fungi applies to the algae. Substantially fewer species of algae have been analyzed biochemically for peroxisome content when compared to the number of ultrastructural examinations (cf. Tables 2.11 and 2.13). Cell fractionation has been applied successfully only to green algal species and *Euglena*. Catalase was found in peroxisomes of all the green algal species examined, and either glycolate or urate oxidase was found in four of the six species studied. Again, the lack of biochemical data and the possible variations in growth conditions, stage of development, etc.,

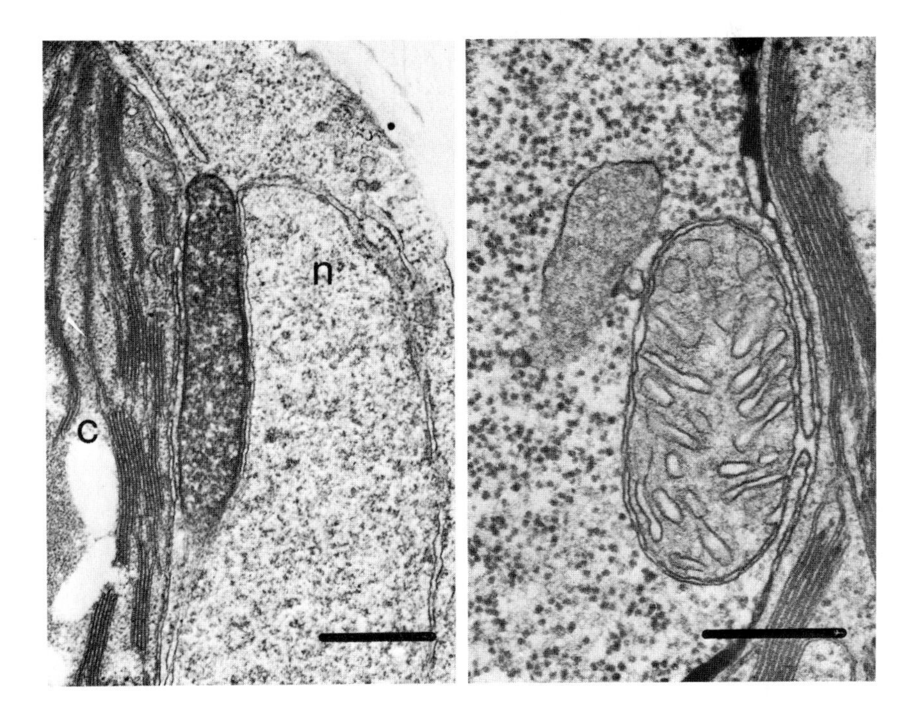

Fig. 2.24. Peroxisomes in green algal cells. (Left) View of the single elongate peroxisome in a filament of *Klebsormidium flaccidum*. The peroxisome, showing catalase reactivity with the DAB reaction, is characteristically sandwiched between the nucleus (n) and chloroplast (c). Bar = 0.5 μm. Micrograph by Nancy Roy. (Right) Profile view of a peroxisome adjacent to a mitochondrion in a *Dunaliella tertiolecta* cell. Bar = 0.5 μm. From S. E. Frederick (unpublished).

TABLE 2.13

Algal Species from Which Peroxisomes Have Been Isolated by Density Gradient Centrifugation[a]

Species	Enzyme activities[b]		
	CAT	GO	UO
Chlorophyta			
Chlamydomonas reinhardii (agarose)	+		
Chlorella vulgaris (acetate/glucose)	+	+	+
Chlorella vulgaris (autotrophic)	+	+	
Chlorogonium elongatum (acetate)	+		+
Polytomella caeca (acetate)	+	−	+
Spirogyra sp. (glucose)	+	+	
Spirogyra varians (autotrophic)	+	+	
Mougeotia sp.	+	+	
Euglenophyta			
Euglena gracilis (ethanol/glucose)	−	−	
Euglena gracilis (hexanoate)	−		
Euglena gracilis (autotrophic)	−		

[a] Enzyme activities associated with the microbodies are indicated by +; the − symbol indicates the enzyme activity was not detected. The carbon source on which the algae were grown is indicated. See Chapter 4 for other details.

[b] Enzyme abbreviations: CAT, catalase; GO, glycolate oxidase; UO, urate oxidase.

make it difficult to draw definitive conclusions on the peroxisomal nature of the microbodies in different algal species. The possible existence of glyoxysome-like peroxisomes and the metabolic function of glycolate oxidase, glycolate dehydrogenase, and their attendant enzymes are explored further in Chapter 4.

IV. SUMMARY

Peroxisomes are present in essentially all cells of flowering and nonflowering plants. They are morphologically simple when compared to other cellular organelles. Their size, shape, and types of inclusions within the matrix, however, are as variable as their number per cell and relative proportions to other organelles when one considers their overall occurrence in the plant kingdom. They may be spherical to ovoid, highly pleiomorphic among the lipid bodies in oil seeds, or extremely elongate (2–3 μm) in certain algal and fungal cells. In root cells and other undifferentiated parenchyma cells, they may be as small as 0.1 μm in diameter, difficult to discern ultrastructurally from the surrounding cytosol, whereas in green leaves, soybean nodules, and *Nitella* filaments they are 15- to 20-fold larger in diameter (1.5–1.8 μm). Only one peroxisome per cell is found in some algae, but they are the most numerous membrane-bound organelles in castor bean endosperm cells. A small or large peroxisome may contain no internal structure, whereas both specialized and unspecialized peroxisomes may contain elaborate crystalline structures, in some cases occupying almost the entire internal space. The occurrence of peroxisomes in all groups of plants has been well documented with the electron microscope, but comprehensive cell fractionation studies are conspicuously lacking in the bryophytes, nonflowering tracheophytes, fungi, and algae. DAB cytochemistry has been used widely to demonstrate catalase reactivity in peroxisomes, although modification of the procedure likely will be required to show its reactivity in many of the fungi and algae. Electron cytochemistry of other peroxisomal enzymes has been employed only to a limited extent among the various organisms, but has potential, especially the cerium perhydroxide method for oxidases, for helping define the functions of peroxisomes, particularly among the lower forms.

3

Methods of Isolation and Physical and Chemical Properties

I. INTRODUCTION

Peroxisomes have a simple structure and molecular composition when compared to other organelles. They contain at most some 30 different enzymes, and catalase may occupy as much as 10–25% of the total organelle protein. The organelle is surrounded by a single membrane and has no internal membrane system. Except in the glyoxysomes, no peroxisomal enzyme has been found associated with the boundary membrane. Crystalloid structures, when present, are believed to be proteinaceous. The presence of DNA and RNA has not been demonstrated unequivocally in peroxisomes (Gerhardt and Beevers, 1969; Ching, 1970; Douglass *et al.*, 1973; Osumi and Kazama, 1978). Although peroxisomes in different tissues and species often have varied matrix and crystalloid morphology (Chapter 2), the organelles are essentially similar in their overall physical characteristics.

II. METHODS OF ISOLATION

A. Theoretical Consideration

Most methods of peroxisome isolation employ centrifugation. The only reported method not involving centrifugation was a preparative sucrose gradient electrophoresis (Theimer and Theimer, 1975), but the separation was inferior to that obtained by sucrose gradient centrifugation. Figure 3.1 is a plot of sedimentation rates versus equilibrium densities of peroxisomes and other organelles in sucrose medium. An examination of the sedimentation coefficients shows that peroxisomes overlap with mitochondria and broken chloroplasts, but are nearly an

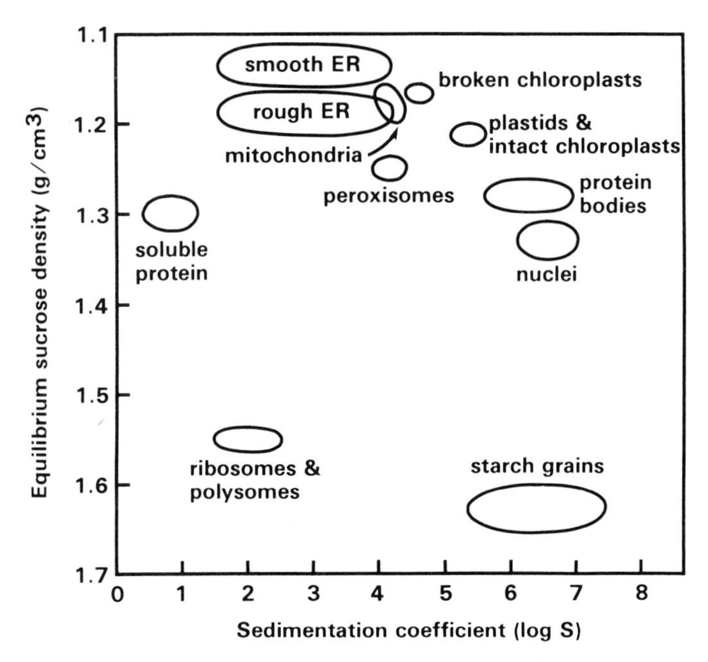

Fig. 3.1. A plot of sedimentation coefficients versus equilibrium densities of plant cell organelles in a sucrose medium. Values are typical for organelles isolated from tissues (storage tissues of oilseeds and leaves) commonly used for cell fractionation studies. Mitochondria from leaves generally have a smaller sedimentation coefficient and a higher equilibrium density. Since the maximal density of a sucrose solution is about 1.4 g/cm^3, the equilibrium densities of ribosomes, starch grains, soluble proteins, and nuclei are those obtained using other density components (Anderson, 1966). The overlapping between rough endoplasmic reticulum and mitochondria has been avoided successfully by omitting Mg^{2+} in the grinding and gradient media such that the rough endoplasmic reticulum is converted artificially to smooth endoplasmic reticulum.

order of magnitude smaller than intact chloroplasts, protein bodies, and nuclei. These relationships indicate that it is possible to enrich intact chloroplasts, protein bodies, and nuclei in one fraction, and peroxisomes, mitochondria, and broken chloroplasts in another fraction by differential (rate) centrifugation. Although the information given in Fig. 3.1 is generally applicable to most plant tissues, preliminary rate centrifugation tests should be done when fractionating tissues not previously studied. Given a pellet composed mostly of mitochondria, peroxisomes, and broken chloroplasts, these organelles can be separated from one another on the basis of their equilibrium densities in sucrose solutions (compare ordinate values in Fig. 3.1). Thus, through a combination of rate and equilibrium density centrifugation, preparation of nearly pure peroxisomes can be successfully obtained from a variety of plant tissues. A more detailed description of the procedures is given below.

Differential centrifugation, as stated above, separates organelles based on their sedimentation coefficients (Fig. 3.1). Centrifuging a crude homogenate at a low speed (e.g., 500 *g*) for several minutes sediments cell debris and large organelles such as nuclei, starch grains, protein bodies, and intact plastids. Centrifugation of the resultant supernatant at 10,000 *g* for approximately 30 min yields a pellet traditionally called the "particulate" or "mitochondrial" fraction which contains most of the peroxisomes, mitochondria, and broken plastids. Although intact plastids and endoplasmic reticulum have sedimentation coefficients higher and lower, respectively, than those of the peroxisomes and mitochondria, a portion of them also sediments in the particulate fraction. When working with leaves, the presence of chloroplasts in the particulate fraction can be reduced greatly by centrifuging the crude homogenate or the 500 *g* supernatant at an intermediate force (e.g., 2000 *g*) before sedimenting mitochondria and peroxisomes at 10,000 *g*. The supernatant from the 10,000 *g* centrifugation can be further centrifuged at 100,000 *g* for approximately 1 hr to pellet "microsomes" containing rough endoplasmic reticulum, ribosomes, and smooth-membrane vesicles originating from the plasma membrane, tonoplast, Golgi, etc. Soluble proteins and other small cellular components remain in the 100,000 *g* supernatant.

Peroxisomes characteristically have a higher equilibrium density in sucrose than many other organelles (Fig. 3.1), and therefore can be separated by isopycnic (equilibrium) centrifugation. Storage protein bodies and plastids have equilibrium densities that overlap with that of peroxisomes; thus, those protein bodies and plastids not removed by differential centrifugation prior to sucrose gradient centrifugation may equilibrate with peroxisomes as contaminants of the fraction. It is noteworthy to mention that the "equilibrium" achieved in a sucrose gradient after centrifugation for a few hours refers to large organelles such as peroxisomes and mitochondria, not small cellular components such as protein molecules and ribosomes. Proteins, nucleic acids, and ribosomes require a long period of ultracentrifugation before their equilibrium densities are reached; in a typical sucrose gradient after several hours of centrifugation, they remain near the top of the sucrose gradient.

B. Grinding Medium and Isolation Procedures

The grinding medium that has been used most extensively for peroxisome isolation is a modification of the first one used to isolate glyoxysomes from castor bean (Breidenbach and Beevers, 1967). The medium contains a buffer (pH 6.9–7.5), 0.4–0.6 *M* sucrose, a sulfhydryl reducing compound, and K^+, Mg^{2+}, and EDTA at low concentrations. For many tissues, the buffer and sucrose appear to be the only essential components. Depending on personal preference and individual tissues, additional components such as bovine serum albumin, strong reducing agents, and Polyclar AT also are used. The volume of grinding medium is normally 2–4 ml per g tissue.

Homogenization can be accomplished by mincing the tissue with medium in an onion chopper and then grinding with a mortar and pestle. A more gentle homogenization procedure involves manual chopping of the tissue with a razor blade in a Petri dish containing grinding medium. A multiple razor blade chopping device (Feierabend and Beevers, 1972b) and two razor blades attached to a modified electric knife (Trelease *et al.*, 1971) also have been used effectively. Generally speaking, razor blade chopping gives much better preservation of peroxisomes, but the yield of organelles is reduced compared to grinding in a mortar. Gentle rupture of protoplasts prepared from spinach leaves and castor bean endosperm has not provided better preservation of intact peroxisomes (Nishimura *et al.*, 1976; Nishimura and Beevers, 1978), although it may be advantageous for other tissues. The homogenate is usually filtered through several layers of medium-moistened cheesecloth or Miracloth, or a layer of Nitex cloth with a defined pore size (e.g., 20 μm^2), to remove cell debris prior to centrifugation.

C. Differential and Gradient Centrifugation

In a typical preparation, the filtered homogenate is centrifuged at the *g* forces described earlier to yield a particulate fraction enriched with mitochondria and peroxisomes prior to sucrose density gradient centifugation. The particulate fraction is resuspended in a few milliliters of grinding medium gently to avoid breakage of the fragile organelles. Resuspension can be achieved by dispersion of the pellet with a rubber policeman followed by repeatedly and slowly drawing the suspension into a 1-ml serological or 5½-in. Pasteur pipette.

Differential centrifugation and resuspension unavoidably leads to breakage and artificial clumping of organelles. To overcome this problem, the homogenate after filtration and/or after an initial low speed centrifugation (e.g., 270–500 *g*, 10 min) is applied directly onto a sucrose density gradient (approximately 5-ml sample on a 30-ml gradient). The tissue must contain a sufficient concentration of organelles such that the 5-ml sample contains enough peroxisomes to be separated on the gradient. Following a standard ultracentrifugation in a swing-out rotor at 65,000 *g* (average) for 4 hr, soluble cellular components and constituents released from organelles remain near the top of the sucrose gradient, and therefore do not interfere with organelle separation within the gradient. Only storage tissues of oilseeds and dark-green leaves (rich in chloroplasts and other organelles) such as spinach leaves have been used successfully for peroxisome isolation from supernatants of low-speed contrifugations. For other tissues which do not contain abundant peroxisomes, the particulate fraction (10,000 *g*) should be prepared first as a mean of concentrating the organelles for subsequent gradient centrifugation.

The density gradient usually is composed of sucrose. Replacing sucrose partly or completely with a sucrose polymer (Ficoll) as a way to reduce osmotic effects may yield better (Parish, 1971) or worse organelle separations (Tolbert, 1971). Silica gel (Percoll) also can be used as a density component, but the separations thus far have not been as good as with sucrose (See Section III). Following equilibrium sucrose gradient centrifugation of preparations from castor bean endosperm or spinach leaves, peroxisomes exhibit a higher equilibrium density than that of mitochondria (Fig. 3.1), and appear as a distinct band below the mitochondrial band (Figure 3.2). The analysis of enzyme distribution on such a gradient from castor bean is shown in Fig. 3.3B. In preparations from other tissues where the peroxisomes are not as abundant as the mitochondria and have an equilibrium density only slightly different from that of the mitochondria, a distinct peroxisomal band is not visible. Plastids, peroxisomes, and mitochondria also can be separated from one another by velocity (rate) density gradient centrifugation where the sedimentation rates of organelles depend on their respective sedimentation

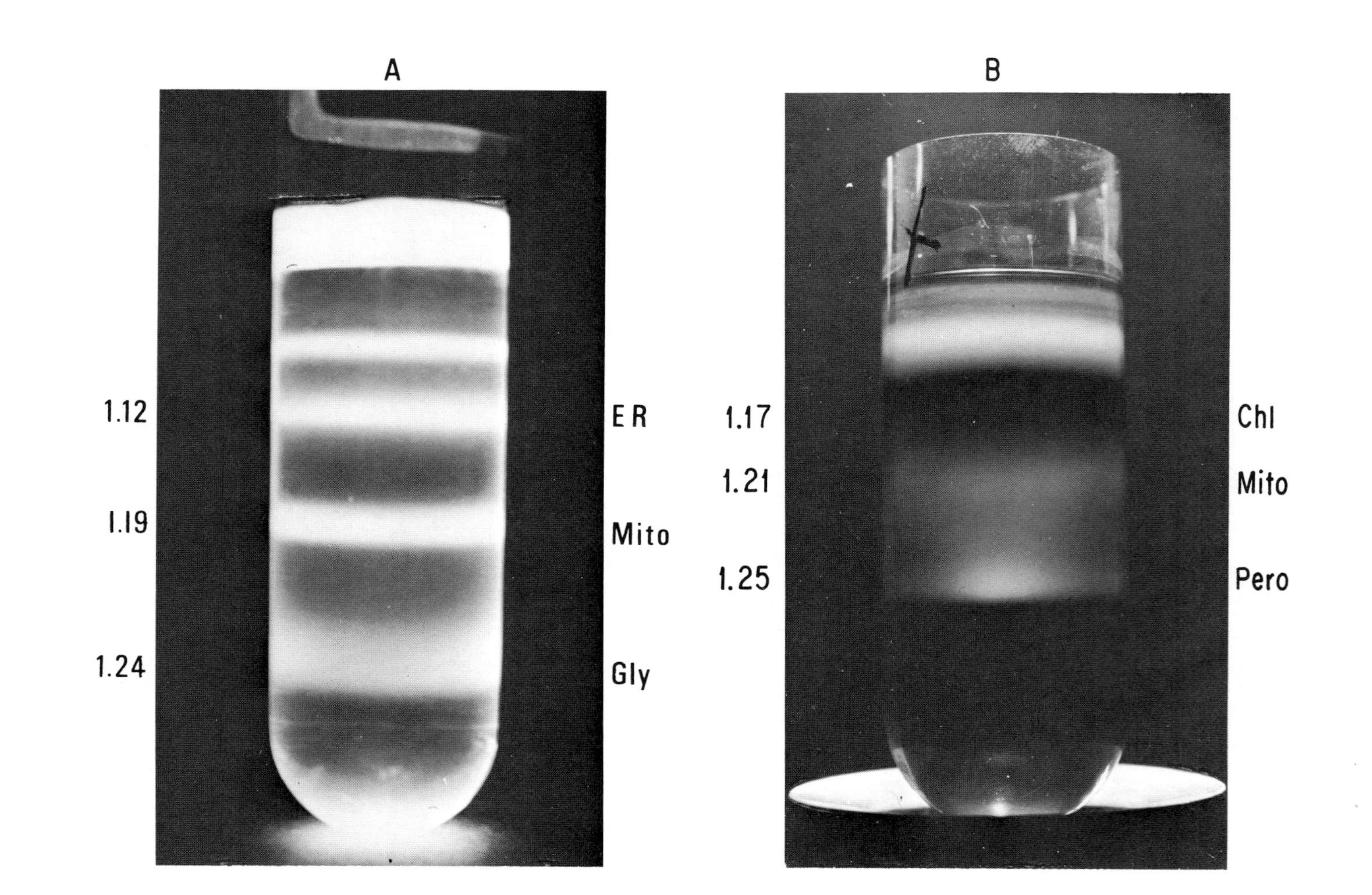
A
1.12
ER
1.19
Mito
1.24
Gly
B
1.17
Chl
1.21
Mito
1.25
Pero

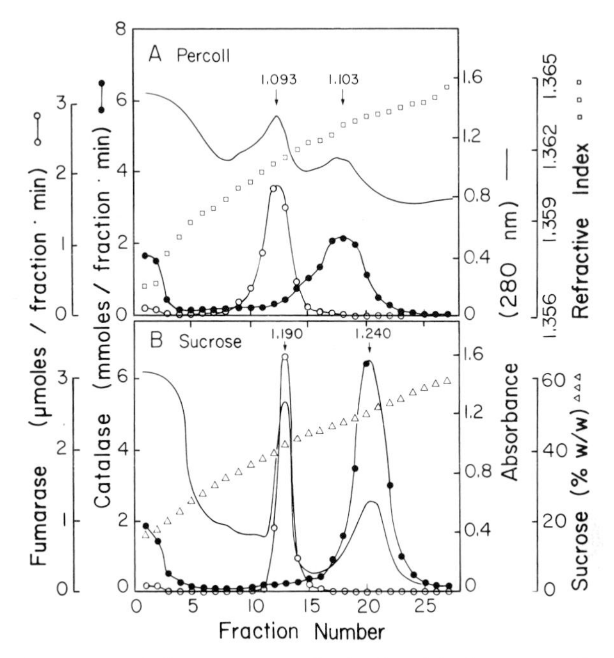

Fig. 3.3. Separation of glyoxysomes and mitochondria from castor bean extract by gradient centrifugation. Catalase and fumarase are markers of glyoxysomes and mitochondria, respectively. After centrifugation of the homogenate at 150 *g* for 10 min to remove plastids and cell debris, the supernatant in 0.45 *M* sucrose was centrifuged in either a 0–45% Percoll gradient containing 0.5 *M* sucrose throughout (A) or a 15–60% (0.45–2.26 *M*) sucrose gradient (B). From Mettler and Beevers (1980).

coefficients as well as equilibrium densities (Fig. 3.1). A comparison of spinach leaf organelle separation in an equilibrium and a velocity sucrose gradient centifugation is shown in Fig. 3.4.

Several modifications of the general procedure have been reported for peroxisome isolation. Zonal (Trelease *et al.*, 1971; Douglass *et al.*, 1973) and vertical rotors (Gregor, 1977) have been used successfully to increase yield and reduce centrifugation time and force, respectively. Dis-

---

Fig. 3.2. Photograph of organelle separation in sucrose density gradients. (A) The homogenate of endosperm extract of 5-day-old castor bean seedlings was centrifuged at 270 *g* for 10 min and the supernatant fraction was applied directly to the gradient. From T. C. Moore (unpublished). (B) Spinach leaf particulate fraction (organelles sedimented at 10,000 *g* for 10 min from a supernatant obtained after centrifugation of the homogenate at 2000 *g* for 10 min) was used. From A. H. C. Huang (unpublished). Scales on left indicate approximate sucrose densities. ER, endoplasmic reticulum; Mito, mitochondria; Gly, glyoxysomes; Chl, broken chloroplasts; Pero, peroxisomes.

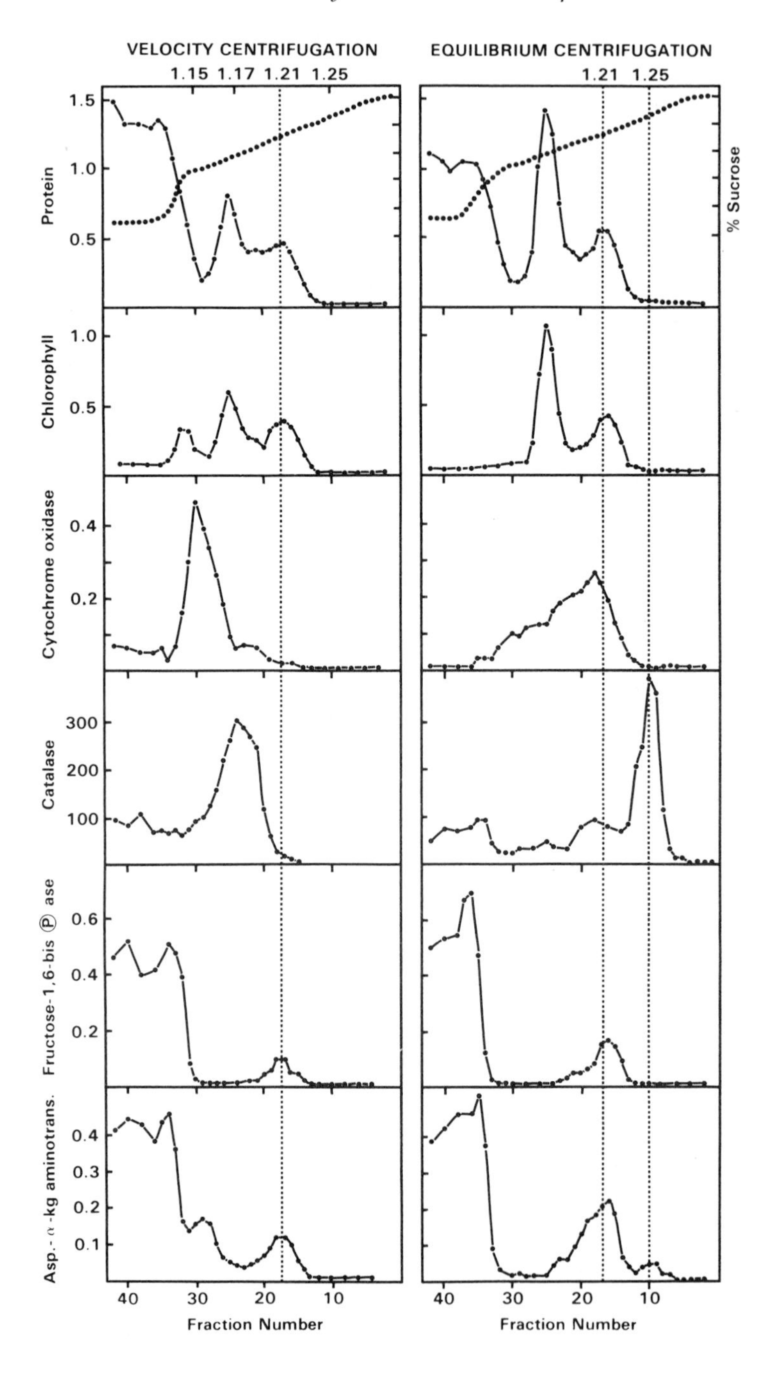

VELOCITY CENTRIFUGATION
EQUILIBRIUM CENTRIFUGATION
1.15 1.17 1.21 1.25
1.21 1.25
Protein
% Sucrose
Chlorophyll
Cytochrome oxidase
Catalase
Fructose-1,6-bis Ⓟ ase
Asp.-α-kg aminotrans.
Fraction Number
Fraction Number

continuous sucrose gradients are employed quite commonly; however, purity of organelles is sacrificed. Flotation gradient centrifugation also has been used to further purify peroxisomes obtained from normal sucrose gradients (Gerhardt and Beever, 1969; Ludwig and Kindl, 1976).

D. Electron Microscopy of Isolated Peroxisomes

The purity and integrity of peroxisomal fractions collected from sucrose gradients can be analyzed with the electron microscope. Special care must be employed during the initial fixation. Usually, the fraction is fixed with glutaraldehyde (2–5%) in an iso-osmotic sucrose solution, i.e., the sucrose concentration of the collected fraction (53–54% w/w). This is accomplished by starting with a high concentration of a stock glutaraldehyde solution (25–70%) and diluting it with a sucrose solution to yield twice the desired final glutaraldehyde concentration in 54% sucrose. This fixative solution is then added to an equal volume of the collected fraction (in 54% sucrose) for 1 to several hours. Thereafter, the fraction is diluted with buffer or lower percent sucrose solutions to approximately 50% sucrose, and the organelles are sedimented by differential centrifugation. The sucrose and glutaldehyde are removed prior to postfixation, dehydration, and embedding in plastic. A common method is to wash the glutaraldehyde-fixed organelles in buffer several times before postfixation in 1–2% osmium tetroxide (Burke and Trelease, 1975). Alternatively, the glutaraldehyde-fixed pellet is washed in buffered 50% sucrose to remove excess glutaraldehyde, then postfixed in 1–2% osmium tetroxide containing buffer and sucrose (Huang and Beevers, 1973). After osmication, the sucrose is removed in water washes. Both methods have been used successfully. Removal of sucrose prior to osmication is simpler and probably has no deleterious effect on organelle integrity.

A main objective of examining gradient fractions with the electron microscope is to analyze all the components in the fractions. Since each gradient fraction contains far too many organelles to be viewed in a few thin sections, steps should be taken to prepare a representative portion of that fraction for assessment in thin sections. Unfortunately, this has

Fig. 3.4. A comparison of organelle separations from spinach leaf extract between a velocity and an equilibrium sucrose density gradient centrifugation. The total leaf extract after filtration was applied directly on sucrose gradients and centrifuged at 12,000 *g* (average) for 10 min (velocity centrifugation) or at 54,000 *g* (average) for 4 hr (equilibrium centrifugation). Cytochrome oxidase (mitochondria), catalase (peroxisomes), and alkaline fructose 1,6-bisphosphatase (intact chloroplasts) were used as markers. From Huang *et al.* (1976).

not been the common practice. Usually, an electron micrograph of the peroxisomal fraction is presented without any explanation of the sampling. Such an electron micrograph may not be a true representation of the whole peroxisomal fraction, since stratification of organelles in the pellet after glutaldehyde fixation might have occurred, and it is difficult to obtain sections of the plastic-embedded pellets along the centrifugal axis. Procedures have been established to view a representative portion of the peroxisomal fraction. For mammalian preparations, Baudhuin *et al.* (1967) modified a Filterfuge tube to collect a representative portion of the fractions on a Millipore filter. A simpler technique accomplishing the same objective has been used for evaluating peroxisomal fractions from plants (Burke and Trelease, 1975). Pellets of glutaraldehyde-fixed organelles are gently and uniformly resuspended in buffer and concentrated on a 0.22 μm cellulose triacetate filter using a Swinney adaptor attached to a syringe. The filters, with a representative portion of the fraction on its surface, are folded once to prevent loss of material, and then encased in 2% warm agar. The fraction can then be washed and postfixed in 1–2% osmium tetroxide. If desired, the fraction can be treated for cytochemical reactivity before osmication. During subsequent dehydration and embedding, the filter dissolves in acetone and monomer plastics, leaving a "line" of organelles in the polymerized plastic. A sectional view perpendicular to the line of organelles reveals the relative proportions of organelles, fragments, etc., of the original peroxisomal fraction (Fig. 3.5). Although the procedure can yield representative sampling, it is still unknown if fragments of organelles observed were derived from intact organelles in the original isolated peroxisomal fraction but were subsequently broken during the electron microscopic procedure.

Figures 3.6 and 3.7 are electron micrographs prepared by sampling procedure often described in the literature of glyoxysomal fractions prepared from oil-rich cucumber cotyledons and castor bean endosperm, respectively. Numerous intact, spherical glyoxysomes are prevalent in the cucumber fraction, but identifiable plastids (differentiating etioplasts) are apparent among the glyoxysomes as are a few intact mitochondria and pieces of disrupted protein bodies. Other unidentified cellular components are present in the fraction, but are minor contaminants. The micrograph illustrating the glyoxysomal fraction isolated from castor-bean endosperm (Fig. 3.7) shows abundant intact glyoxysomes essentially free of contaminating organelles. Preparations of peroxisomes from castor bean are the most pure fractions that have been made from any plant tissue, and in this sense may be unique to castor

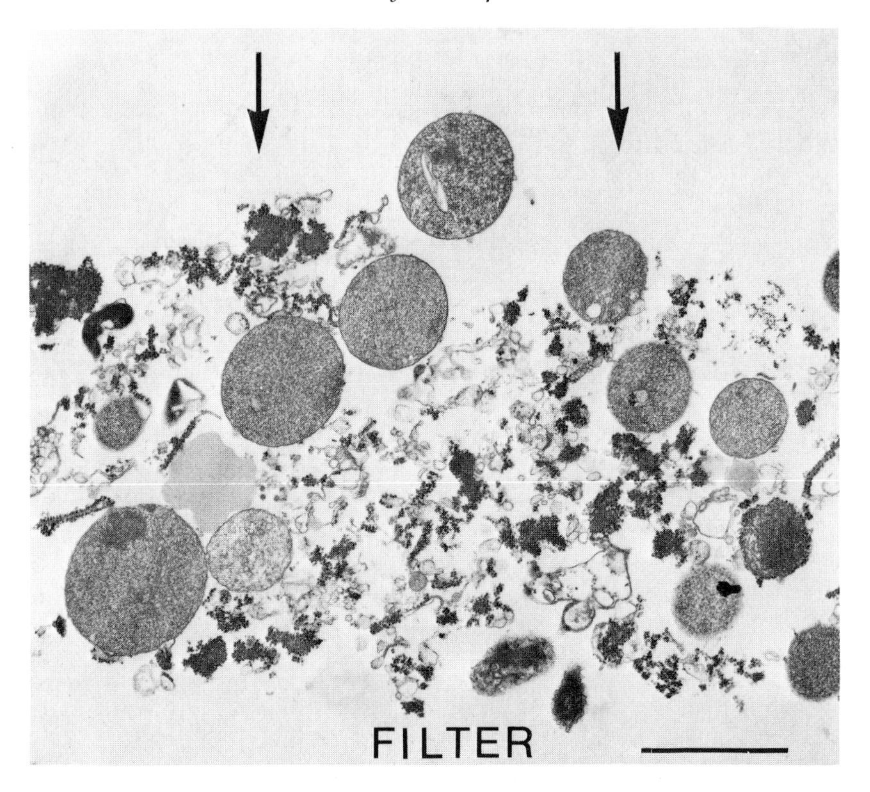

Fig. 3.5. Electron micrograph of a portion of the cottonseed glyoxysomal fraction isolated by sucrose gradient centrifugation. A filtration procedure was used to obtain a representative sample of the whole gradient fraction for assessment in thin electron microscopic sections. The glyoxysomes were the major intact organelles on top of the filter which had been dissolved during the dehydration and embedding steps. Arrow indicates the direction of filtration. Bar = 1 μm. From R. N. Trelease (unpublished).

bean. It is evident from numerous studies that highly enriched preparations of mostly intact peroxisomes can be isolated from plants.

III. PHYSICAL PROPERTIES

In sucrose gradient centrifugations, the equilibrium densities of peroxisomes from different tissues range from 1.20 to 1.26 g/cm^3 (Gerhardt, 1978). Equilibrium densities of less than 1.20 g/cm^3 occasionally have been reported. It is assumed that the peroxisomal membrane is permeable to sucrose. During gradient centrifugation, sucrose gradually enters

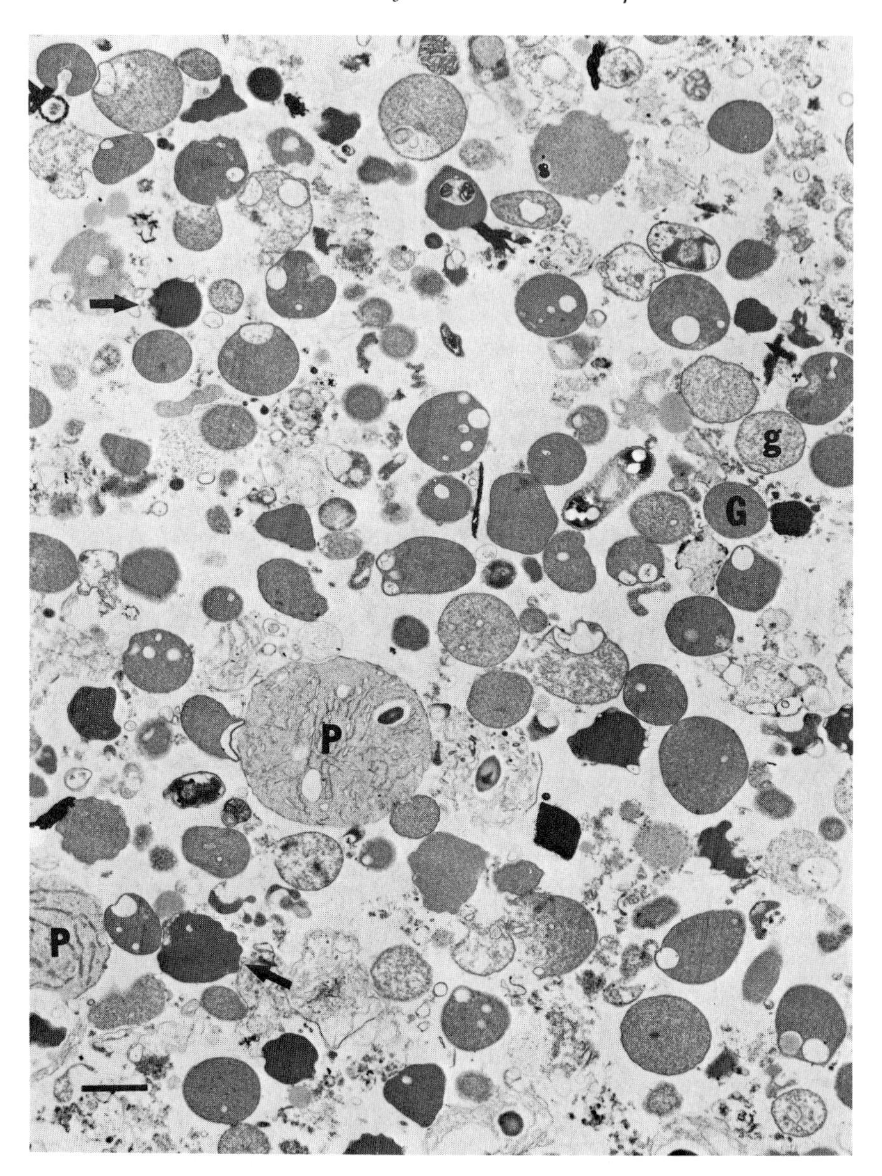

Fig. 3.6. Electron micrograph of the cucumber glyoxysomal fraction isolated by sucrose gradient centrifugation. Intact glyoxysomes (G) and glyoxysomes with their content partially deleted (g) are the dominant organelles, whereas plastids (P) and protein body fragments (arrows) are visible contaminants. Bar = 1 μm. From J. J. Burke (unpublished).

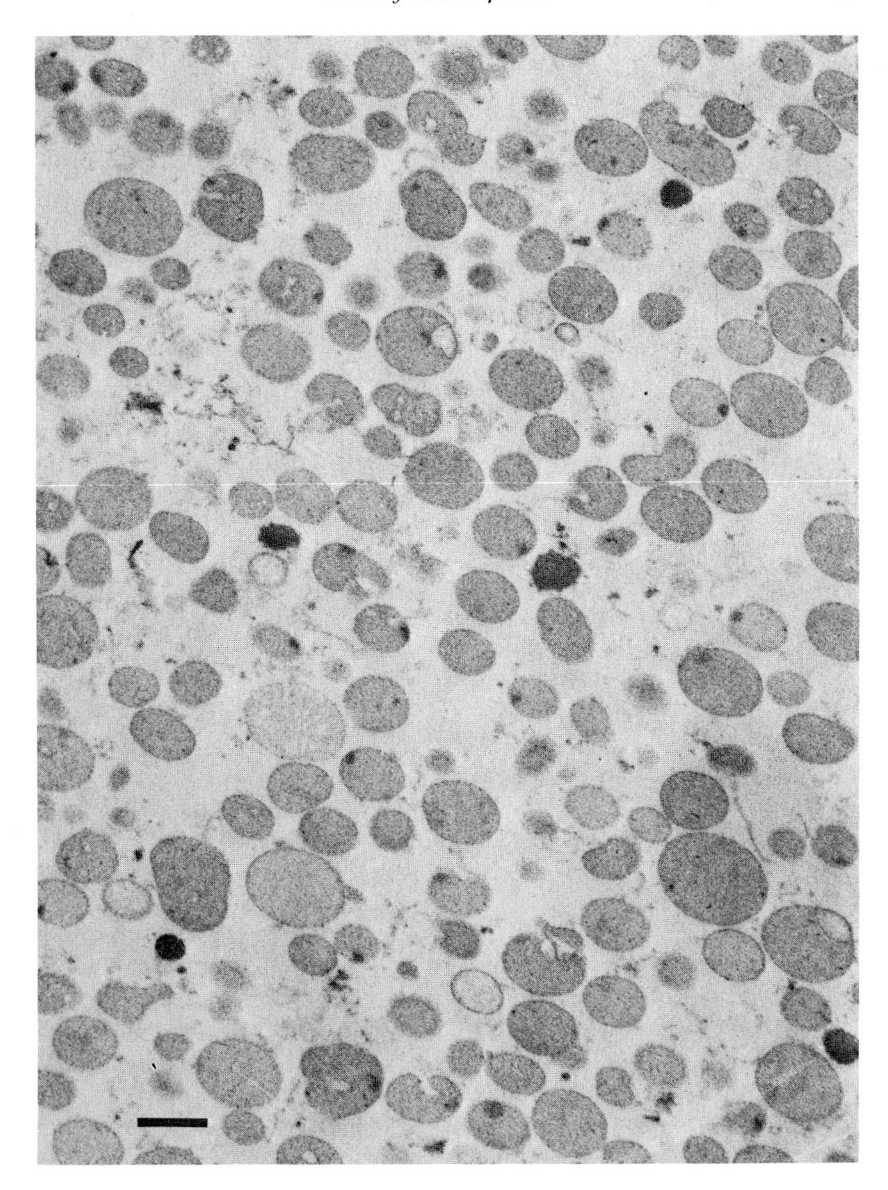

Fig. 3.7. Electron micrograph of the castor bean glyoxysomal fraction isolated by sucrose gradient centrifugation. Bar = 1 μm.

the organelle such that the matrix solution ultimately becomes equilibrated with the external solution. As a consequence, the equilibrium density of a peroxisome depends largely on the density of its molecular components, mostly protein and membrane lipid. Therefore, peroxisomes have an equilibrium density lower than that of soluble proteins but higher than those of the double-membraned mitochondria and plastids (Fig. 3.1). In some density gradient centrifugations, sucrose is replaced by a high molecular weight compound (e.g., Ficoll, Percoll) to minimize osmotic effects. In these preparations, the high molecular weight compounds cannot enter the peroxisomes, and thus the equilibrium density of the peroxisomes is less than that in sucrose and becomes closer to that of the chloroplasts or mitochondria (Tolbert, 1971; Mettler and Beevers, 1980; Liang *et al.*, 1982). A comparison between sucrose and Percoll gradient centrifugation of organelles from castor bean extract is illustrated in Fig. 3.3.

When the peroxisomal fraction taken from the 1.25 g/cm^3 region of a sucrose gradient is diluted with water or with a less concentrated sucrose solution, the rate of sucrose molecules leaving the organelles is much lower than that of water molecules entering the organelles due to osmosis. The organelles rapidly increase their volume and finally burst. Attempts to reduce the sucrose concentration of an isolated peroxisome fraction without lyzing most of the organelles have not been successful. This failure has prevented the studies of peroxisomal metabolism and biogenesis under *in vivo* conditions. The recently established procedure for preparing peroxisomes in dilute osmotica by Percoll density gradient centrifugation would make these studies possible (Mettler and Beevers, 1980; Liang *et al.*, 1982; Neuburger *et al.*, 1982; Schmitt and Edwards, 1982).

It is still unknown why the equilibrium densities of peroxisomes from different tissues exhibit such a variable range (1.20–1.26 g/cm^3). This may reflect diverse properties, contents, and/or physiological functions of peroxisomes in the various tissues. In some cases, it may represent an artificial consequence of the preparation procedure during which certain compounds leak from the organelles. Such leakage would be tissue-specific, depending on the cellular conditions such as the stiffness of the cell wall, and the amount of acid and other interfering compounds. Both glyoxysomes and leaf peroxisomes have a similar equilibrium density of 1.25–1.26 g/cm^3, irrespective of the species from which they are isolated. An interesting exception is the glyoxysomes from castor bean. The castor bean glyoxysomes had an equilibrium density of 1.25 g/cm^3 when they were first isolated (Breidenbach and Beevers, 1967), but similar experiments done in later years using other batches of the same seed variety indicated that the glyoxysomes had an equilibrium density of 1.23 g/cm^3. The reason for this difference in the equilibrium density between the original and the subsequent reports is not understood, as is the reason for the diversion in the equilibrium density of the castor bean glyoxysomes from the glyoxysomes of other oilseeds. In castor bean, contamination of the glyoxysomal fraction by plastids which have an equilibrium density of 1.22 g/cm^3 has become a problem. Investigators should take special precautions with the castor bean glyoxysome fractions when studying certain aspects of the organelles, such as anlyzing the membrane lipid components.

Little is known about the physical properties of plant peroxisomes other than their behavior during sucrose gradient centrifugation. The known physical properties of mammalian liver peroxisomes are shown in Table 3.1, but insufficient information on plant peroxisomes is available for comparison. Liver peroxisomes have a sedimentation coefficient half that of mitochondria in 0.25 *M* sucrose. On the contrary, spinach leaf (Fig. 3.4) and castor bean (Miflin and Beevers, 1974) peroxisomes migrate faster than the mitochondria during velocity sucrose gradient centrifugation. Also, spinach leaf peroxisomes migrate faster than mitochondria in a Percoll density gradient containing 0.4 *M* sucrose (Liang *et al.*, 1982). In 0.25 *M* sucrose, both spinach peroxisomes and rat liver peroxisomes have an equilibrium density of 1.09 g/cm^3.

TABLE 3.1

Physical Properties of Liver Peroxisomes[a]

Parameter	Peroxisomes	Mitochondria
Dry weight (10^{-14} g)	2.4	10
Dry density	1.32	1.32
Osmotically active solutes (mosmols/g dry wt)	0	0.16
Water compartments (cm^3/g dry wt)		
Hydration	0.27	0.43
Sucrose space	2.58	0.91
Osmotic space in 0.25 *M* sucrose	0	0.60
Total in 0.25 *M* sucrose	2.85	1.93
Sedimentation coefficient in 0.25 *M* sucrose (10^3 Svedberg units)	4.4	10
Diameter in 0.25 *M* sucrose (10^{-5} cm)	5.4	8
Density in 0.25 *M* sucrose	1.09	1.10
Volume in 0.25 *M* sucrose (10^{-14} cm^3)	8.2	27

[a] Information on liver mitochondria are provided for comparison. Values of peroxisomes are the average of those calculated separately using urate oxidase, catalase, and D-amino acid oxidase. From de Duve and Baudhuin (1966).

IV. SUBFRACTIONATION AND CHEMICAL PROPERTIES

Peroxisomes have been subfractionated into their membrane and matrix. The peroxisomes in most plant tissues do not contain crystalloids (Chapter 2). The crystalloid components have been subfractionated only from potato tuber peroxisomes. Although the leaf tissues of some species (e.g., tobacco) are known to contain abundant and large crystalloids in the peroxisomes (Frederick and Newcomb, 1969a,b), no biochemical study on these crystalloids has been made. Extensive subfractionation studies have been performed with castor bean glyoxysomes (Bieglmayer *et al.*, 1973; Huang and Beevers, 1973), and similar findings on the glyoxysomes from numerous other oil seeds have been reported (Gerhardt, 1978).

Glyoxysomal fraction isolated in a sucrose density gradient is lysed osmotically by dilution with buffer or water. Subsequently, the membranes are pelleted by high speed centrifugation, leaving the solubilized matrix proteins in the supernatant. The membrane pellet contains glyoxysomal ghosts; those from castor bean are shown in Fig. 3.8. Washing the membranes with 0.2 *M* KCl did not affect the ultrastructural appearance of the ghosts even though some specific proteins were removed (see Fig. 4.5). When glyoxysomal ghosts were centrifuged on a sucrose gradient, they reached an equilibrium density of 1.21 g/cm^3. The decrease in equilibrium density presumably reflects the loss of matrix proteins (equilibrium density $\simeq$ 1.30 g/cm^3). Although many glyoxysomal enzymes are associated with the boundary membrane to varying degrees, more than 50% of the organelle protein is matrix protein (Chapter 4).

The lipid and protein components of the salt-washed castor bean glyoxysomal ghosts have been examined. Phosphatidylcholine (51%) and phosphatidylethanolamine (27%) were the major phospholipid components (Donaldson *et al.*, 1972). When salt-washed glyoxysomal ghosts from pumpkin and castor beans were subjected to sodium dodecyl sulfate–acrylamide gel electrophoresis (Bieglmayer and Ruis, 1974; Brown *et al.*, 1974), more than 10 protein bands of molecular weight ranging from 13,000 to 97,000 were observed (Fig. 3.9). One prominent band had a molecular weight of 63,000–66,000. Unlike the minor proteins, this protein resisted solubilization by cholate, deoxycholate, KCl, Triton X-100, and chloroform/methanol; it was solubilized by SDS or guanidinium chloride. The protein apparently is not a glycoprotein, and has been observed in glyoxysomes isolated from castor bean, cucumber, sunflower, and pumpkin. It is also the major protein component of leaf peroxisomes from *Lens culinaris* and green pumpkin cotyledons (Brown

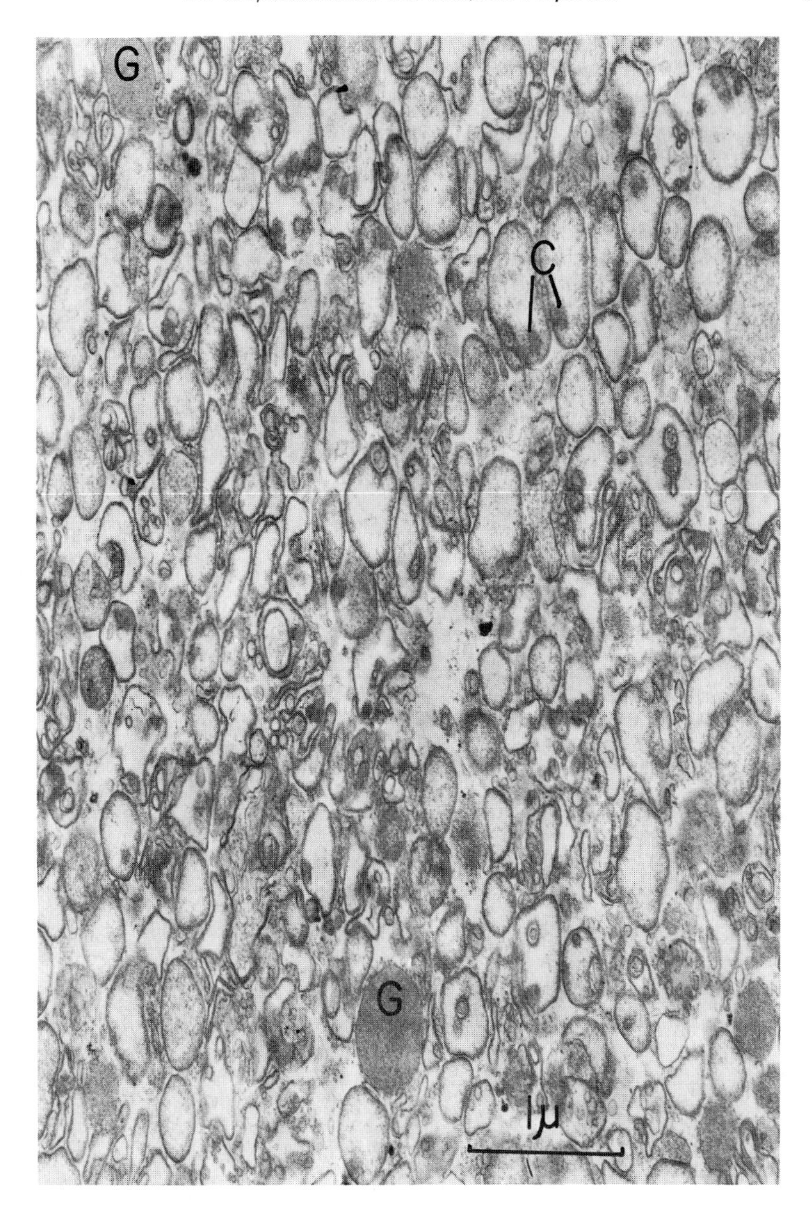

Fig. 3.8. Electron micrograph of the castor bean glyoxysomal fraction after osmotic shock. The membranes appear intact and the cores (C), where present, occur within the ghosts. A few of the glyoxysomes (G) still retained their amorphous matrix. Bar = 1 μm. From Huang and Beevers (1973).

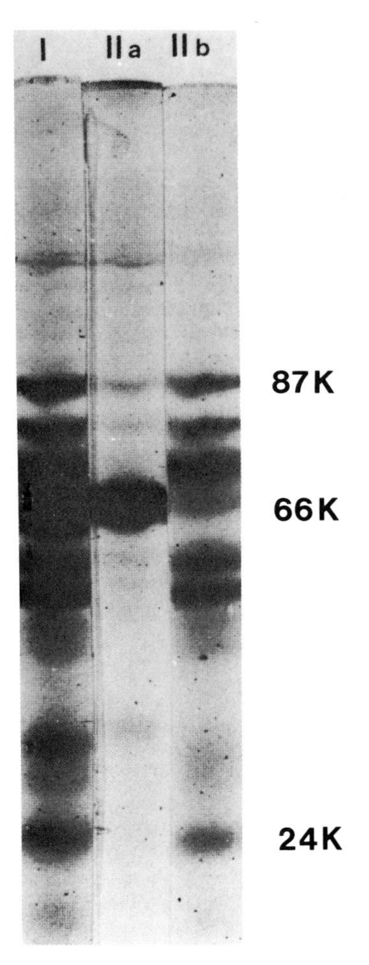

Fig. 3.9. Sodium dodecyl sulfate–polyacrylamide gel electrophoresis of membrane proteins of castor bean glyoxysomes. The membrane ghosts (I) were treated with 1% sodium deoxycholate, and centrifuged at 100,000 *g* for 30 min to yield a pellet (IIa) and a supernatant (IIb). The numbers on the right-hand side are estimated molecular weights. From Bieglmayer and Ruis (1974).

et al., 1974; Ludwig and Kindl, 1976). Whereas this protein was reported to be absent from the plastids, mitochondria, and endoplasmic reticulum of *Lens culinaris* leaves (Ludwig and Kindl, 1976), it was found in the endoplasmic reticulum and the outer mitochondrial membrane of castor bean (Bieglmayer and Ruis, 1974). On the basis of its prominence and its resistance to solubilization, the protein has been suggested to be the major structural protein of the peroxisomal membrane.

Approximately 60–70% of the glyoxysomal protein is present in the matrix. Undoubtedly, a substantial portion of the matrix protein is catalase. Although many of the β-oxidation and glyoxylate cycle enzymes are membrane associated, they can be solubilized from the membrane by mild salt treatment (Chapter 4). The solubilized proteins after osmotic and salt treatment can be resolved into several protein bands by sodium dodecyl sulfate–polyacrylamide gel electrophoresis (Mellor *et al.*, 1978). Some of the minor protein bands show positive glycoprotein stain with periodic acid–Schiff's reagent. However, no peroxisomal enzyme has been identified to be a glycoprotein, and the possibility of the glycoprotein bands being contaminants (especially from the plastids) has not been eliminated (Bergner and Tanner, 1981; Lord and Roberts, 1982).

Peroxisomes other than glyoxysomes do not appear to have enzymes associated with the membrane. Similar osmotic shock treatments were performed on peroxisomes isolated from spinach leaves and potato tuber (as a representative of the unspecialized peroxisomes). Most of the enzymes examined were solubilized, not associated with the membrane pellet (see Table 4.5). The phospholipid content of peroxisomes other than glyoxysomes has not been reported.

The peroxisomes from castor bean endosperm and spinach leaves are generally devoid of protein crystalloids. Most of the peroxisomes isolated from potato tuber contain protein crystalloids (Neuburger *et al.*, 1982). These protein crystalloids have been isolated by sucrose gradient centrifugation (Huang and Beevers, 1973). They have an equilibrium density of 1.28 g/cm^3, similar to the crystalloids isolated from rat liver peroxisomes (Baudhuin *et al.*, 1965; Hayashi *et al.*, 1976). No urate oxidase or glycolate oxidase, but 7% of the total peroxisomal catalase, was found in the crystalloid fraction. Rat liver crystalloids do not possess catalase, but all of the peroxisomal urate oxidase. Thus, there may be a major difference in enzyme components between the crystalloids of plant (potato tuber) and animal (rat liver) peroxisomes.

V. SUMMARY

Peroxisomes have simple morphological features and molecular constituents. Owing to their high equilibrium density as compared with those of other organelles, peroxisomes have been isolated routinely by sucrose density gradient centrifugation. Subfractionation of peroxisomes has been done almost exclusively on glyoxysomes, mostly those from castor bean endosperm. Glyoxysomes are lyzed osmotically, and the membrane and matrix are separated from each other by centrifugation.

The lipids and proteins of the membrane, as well as the proteins of the matrix, have been characterized. Many of the glyoxylate cycle and β-oxidation enzymes are associated with the membrane to varying degrees, whereas other enzymes, including catalase and urate oxidase, are in the matrix. In peroxisomes other than glyoxysomes, the enzymes that have been examined are present in the matrix, not associated with the membrane. Peroxisomes from most plant tissues do not contain crystalloids; the crystalloids isolated from potato tuber peroxisomes, unlike those of rat liver peroxisomes, contain catalase but not urate oxidase.

4

Metabolism, Enzymology, and Function

I. INTRODUCTION

As mentioned in Chapter 1, peroxisomes from various tissues share among themselves common enzyme constituents, including catalase and one or more H_2O_2-producing oxidases. With few exceptions, the activity of catalase is extremely high (Tables 4.1 and 4.2). On the other hand, the activities of glycolate oxidase and urate oxidase, the two most widely distributed H_2O_2-producing oxidases, are comparatively low. Furthermore, the activities of these two enzymes vary greatly among peroxisomes of diverse tissues. The variation reflects the relative importance of the respective oxidases, and thus of the peroxisomes in the tissues where they are located, in performing unique physiological functions. For example, in leaf peroxisomes, which are essential for photorespiration, the activity of glycolate oxidase is much higher than those in peroxisomes of other tissues. In rat liver and symbiotic root nodules, which are active in ureide catabolism, the urate oxidase activity in the peroxisomes is much higher than those in tissues that are not involved

TABLE 4.1

Enzyme Activities in Peroxisomes Isolated from Various Tissues[a]

	Enzyme activity (nmol/min/mg protein)				
	Potato tuber	Castor bean endosperm	Spinach leaves	Soybean nodule	Rat liver
Catalase	0.6×10^6	13×10^6	2.2×10^6	0.16×10^6	11.4×10^6
Glycolate oxidase	34	95	1,130	—[b]	32
Urate oxidase	19	16	10	89	133
Hydroxypyruvate reductase	—	36	6,500	—	—
Malate dehydrogenase	—	38,000	21,600	—	—
Aspartate–α-ketoglutarate aminotransferase	—	4,200	250	—	—
Thiolase	—	280	8	—	—
Enoyl-CoA hydratase	—	19,800	671	—	—
β-OH-acyl-CoA dehydrogenase	—	15,900	32	—	—
Citrate synthase	—	840	—	—	—
Malate synthase	—	2,100	—	—	—
Isocitrate lyase	—	930	—	—	—

[a] Potato tuber peroxisomes (representing the unspecialized peroxisomes), and rat liver peroxisomes are provided for comparison. The results on potato tuber, castor bean endosperm, spinach leaves, and rat liver were obtained using identical organelle preparation procedure and enzyme assay methods (Huang and Beevers, 1973). The presence of low activities of β-oxidation enzymes in spinach peroxisomes was reported by Gerhardt (1981). The activities of catalase and urate oxidase in soybean nodule peroxisomes should be higher since the isolated peroxisomal fraction was contaminated by bacteria (Hanks *et al.*, 1981).

[b] Activities not tested.

in ureide catabolism. Whereas glycolate oxidase and urate oxidase perform well-defined physiological roles in leaf peroxisomes and liver and root nodule peroxisomes, respectively, their function in peroxisomes of other tissues appears to be limited. In these other tissues, the activities of the two oxidases are comparatively low both in the total tissue extracts and in isolated peroxisomes, and there is no known source of substrates that would be utilized under physiological conditions. The ubiquitous occurrence of the H_2O_2-producing oxidases, as well as catalase, may simply reflect the basic and common enzyme constituents of peroxisomes that have the potential to become specialized for unique functions by the addition of extra enzymes as well as enzymes that were not present.

TABLE 4.2

Enzyme Activities in Isolated Castor Bean Glyoxysomes and Spinach Leaf Peroxisomes[a]

Enzymes	Enzyme (μmol/min/mg protein)	
	Castor bean glyoxysomes	Spinach leaf peroxisomes
Catalase	5,000–13,000	2,400–12,000
Glycolate oxidase	0.07–0.1	1–10
Allantoinase	0.002–0.01	
Urate oxidase	0.02	0.01
Hydroxypyruvate reductase	0.04–0.2	7–47
NADH-cytochrome reductase	0.04	0.005–0.02
NADH-ferricyanide reductase	0.8	
Alkaline lipase	0.2	
Fatty acyl-CoA thiokinase	0.1	
Fatty acyl-CoA oxidase	0.2	0.02
Enoyl-CoA hydratase	3–20	0.67
β-Hydroxyacyl-CoA dehydrogenase	3–16	0.03
Thiolase	0.2–1.0	0.008
Acetothiokinase	0.05	
Isocitrate lyase	0.7–2.0	
Malate synthase	1.5–2.1	
Malate dehydrogenase	17–38	22–87
Aconitase	0.07	
Citrate synthase	0.08–1.2	
Aspartate–α-ketoglutarate aminotransferase	2.3–7.7	0.2–7.3
Glutamate–glyoxylate aminotransferase	3.8	2.4–7
Serine–glyoxylate aminotransferase	0.16	1.5–12

[a] The values are compiled from the reviews of Tolbert (1971), Beevers and Breidenbach (1974), and Gerhardt (1978). The enzyme activities were reported by investigators using different organelle isolation procedure and assay methods. Use Table 4.1 for direct comparison. In spinach leaf peroxisomes, great variations in activities exist in the glycolate pathway enzymes; at least for the aminotransferases, the activities are probably on the lower end of the listed variations at 1–2 μmol/min/mg protein (Tolbert, 1980).

II. CHARACTERISTIC PEROXISOMAL ENZYMES

A. Catalase

With few exceptions, peroxisomes isolated from various tissues contain a high catalase activity. Catalase comprises as much as 10–25% of the total peroxisomal proteins (Tolbert, 1980). Its extremely high activity in the peroxisomes is due to its abundance and high catalytic turnover number of about 10^7 molecules of H_2O_2 decomposed per catalase molecule per second at pH 7 and 30°C. The enzyme has been purified from leaves, green cotyledons and yeast (Table 4.3). It has a molecular weight ranging from 225,000 to 300,000 and consists of four identical subunits with heme as its prosthetic group. Its catalysis is as follows (review by Deisseroth and Dounce, 1970):

$$\text{Catalase} + H_2O_2 \rightarrow \text{catalase } H_2O_2 \text{ (complex 1)} \quad (4.1)$$

and either

$$\text{Catalase } H_2O_2 + H_2O_2 \rightarrow \text{catalase} + 2\ H_2O + O_2 \quad \text{(catalatic reaction)} \quad (4.2)$$

or

$$\text{Catalase } H_2O_2 + H_2R \rightarrow \text{catalase} + 2\ H_2O + R \quad \text{(peroxidative reaction)} \quad (4.3)$$

H_2R may be any simple metabolite, such as ethanol, methanol, formaldehyde, or formate, or more complex artificial molecules such as 3,3′-diaminobenzidine.

Catalase is a rather peculiar enzyme in that the apparent K_m value for H_2O_2 is over 1 *M*, a concentration that is unlikely to exist *in vivo.* In most reports, catalase activity was measured by the decrease of H_2O_2 at 240 nm, using a H_2O_2 concentration of approximately 0.014 *M* (Lück, 1965). The *in vitro* specific activity recorded for isolated peroxisomes has been in the range of $1–10 \times 10^3$ μmoles/min/mg protein (Table 4.1). This activity is several orders of magnitude higher than those of other enzymes in the peroxisomes, especially the H_2O_2-producing oxidases. Without knowing the concentration of H_2O_2 *in vivo,* the excess catalase activity over the H_2O_2-producing oxidase activities recorded under *in vitro* condition should be taken with caution. Nevertheless, it is generally assumed that the high activity of catalase would ensure the complete destruction of H_2O_2 as soon as it is produced.

Whether such an assumption is valid is crucial in two aspects of peroxisomal metabolism. The first aspect involves the extent of peroxidative activity of catalase *in vivo.* Catalase preferentially participates in the catalatic reaction under relatively high H_2O_2 concentration and neutral pH, whereas it is active in the peroxidative reaction at a low H_2O_2 concentration and alkaline pH. If the peroxidative activity is important *in vivo,* then it follows that the H_2O_2 concentration in the peroxisomes would be relatively low. This condition is dependent on the activity of the H_2O_2-producing oxidases and the amount of H_2O_2 that could possibly accumulate before being destroyed by catalase. Although the peroxidative activity of catalase in isolated peroxisomes has been demonstrated, the extent of the activity *in vivo* is unknown. Furthermore, specific reductants for the peroxidative activity under physiological conditions have not been invoked (except in mammalian peroxisomes). The second aspect deals with the possible nonenzymatic reaction of H_2O_2 with reactive metabolites such as glyoxylate, both produced in peroxisomes. It is known that glyoxylate is converted to malate in the glyoxysomes (Section IV C) and to glycine in the leaf peroxisomes (Section V B). However, the possibility has been raised that some glyoxylate in leaf peroxisomes may react with the H_2O_2 that has escaped destruction by catalase to produce formate and CO_2. Whether this reaction occurs at a significant level *in vivo* depends on the actual activity of catalase within the peroxisomes.

B. Glycolate Oxidase

Glycolate oxidase catalyzes the following reaction:

$$\underset{\text{Glycolate}}{CH_2OHCOOH} + O_2 \rightarrow \underset{\text{Glyoxylate}}{CHOCOOH} + H_2O_2 \tag{4.4}$$

The enzyme from green leaves has been purified and extensively studied (Table 4.3). The enzyme from the nongreen cotyledons of cucumber seedling also has been characterized (Kindl, 1982). It has a molecular weight of about 225,000–300,000 and consists of four identical subunits each having FMN as the cofactor. The enzyme shares many similar properties with other peroxisomal flavin oxidases such as fatty acyl-CoA oxidase and methanol oxidase. The catalysis occurs optimally at pH 8.5. The enzyme has a low affinity for O_2 ($K_m = 1.3 \times 10^{-4}$ *M*), and saturation occurs only at 60% O_2. Under anaerobic conditions, 2,6-dichlorophenol indophenol and different quinones can act as artificial electron acceptors.

The enzyme catalyzes the oxidation of glycolate as well as other α-hydroxyacids, such as L-lactate, α-hydroxycaproate, and several analogs of indoleglycolate, phenylglycolate, and lactate, and thus also is referred to as α-hydroxyacid oxidase. However, it shows highest V_{max} and affinity ($K_m = 2 \times 10^{-4}$ *M*) toward glycolate. Other α-hydroxyacids that can be inhibitory substrate analogs include α-hydroxysulfonates, isonicotinylhydrazide, and 2-hydroxy-3-butinoate. These inhibitors are very potent; hydroxymethanesulfonate at 5×10^{-5} *M* competitively inhibits 50% of the enzyme activity at a glycolate concentration of 10^{-2} *M* (Zelitch, 1971). They have been used in *in vivo* studies in order to assess the physiological role of glycolate oxidase in photorespiration. The enzyme also can catalyze the oxidation of glyoxylate (the reaction

TABLE 4.3

Properties of Peroxisomal Enzymes[a]

Enzymes	Tissues	Molecular weight (subunit)	Optimal pH for activity	Apparent K_m values (μM)	References
Catalase	Spinach leaf Sunflower cotyledon	300,000 (72,000)	7–8	> 1*M*	Gregory (1968), Gerhardt (1978)
Glycolate oxidase	Spinach leaf Pea leaf	140,000 (70,000)	8	262 (glycolate); 133 (O_2)	Frigerio and Harbury (1958), Kerr and Groves (1975)
Hydroxypryruvate reductase	Spinach leaf	97,500	6.5	50 (OH pyruvate); 50,000 (glyoxylate)	Kohn and Warren (1970)
Urate oxidase	*Candida utilis*	120,000	8.5	10.5	Itaya *et al.* (1971)
Isocitrate lyase	Flax cotyledon Cucumber cotyledon	264,000 (67,000)	7.5	320	Khan *et al.* (1977), Lamb *et al.* (1978)
Malate synthase	Castor bean endosperm Cotton cotyledon	575,000 (64,000)	7.2	52 (glyoxylate); 10 (acetyl-CoA)	Bowden and Lord (1978), Miernyk and Trelease (1981d)
Citrate synthase	Castor bean endosperm Cucumber cotyledon	100,000 (46,000)	7.5	7 (oxaloacetate); 80 (acetyl-CoA)	Huang *et al.* (1974), Köller and Kindl (1977)
Malate dehydrogenase	Spinach leaf	67,000 (33,500)	7.0	39 (oxaloacetate);	Rocha and Ting (1970),

	Cucumber cotyledon			24 (NADH)	Köller and Kindl (1977)
Lipase	Castor bean endosperm	Unknown	9.0	230 (*N*-methyl indoxylmyristate	Muto and Beevers (1974)
β-*OH*-Acyl-CoA dehydrogenase; enoyl-CoA dehydrogenase[b]	Cucumber cotyledon	75,000	Unknown	Unknown	Frevert and Kindl (1980b)
Thiolase	Cucumber cotyledon	90,000 (45,000)	Unknown	Unknown	Frevert and Kindl (1980b)
Serine-glyoxylate aminotransferase	Spinach leaf	Unknown	7.0	2720 (serine) 150 (glyoxylate)	Rehfeld and Tolbert (1972)
Glutamate-glyoxylate aminotransferase	Spinach leaf	Unknown	7.0	3600 (glutamate); 4400 (glyoxylate)	Kisaki and Tolbert (1969)
Monoamine oxidase	*Aspergillus niger*	252,000	7–8	17 (benzlamine)	Yamada *et al.* (1965)
Methanol oxidase	*Candida boidinii*	600,000 (74,000)	(7.5–9.5)	2000 (methanol) 4500 (ethanol)	Sahm and Wagner (1973)

[a] The properties of several enzymes have been reported by many different laboratories. The purpose of this table is to provide an overall view of the properties of peroxisomal enzymes, not as a review of all the references. Only one to two references are provided for each enzyme. The selection of references is partially based on the details of the reports, the representation of similar properties among enzymes from different species, and the preferences on higher plants whenever available.

[b] As a bifunctional protein.

product from glycolate) to oxalate, since glyoxylate occurs as a hydrated form in solution [CH—$(OH)_2$—COOH] and thus is similar structurally to an α-hydroxyacid. The significance of this reaction will be discussed in Section V,B on photorespiration.

C. Urate Oxidase (Uricase)

Urate oxidase catalyzes the following reaction:

$$\text{urate} + O_2 + 2H_2O \rightarrow \text{allantoin} + H_2O_2 + CO_2 \quad (4.5)$$

urate allantoin

The enzyme has been detected in peroxisomes isolated from a wide variety of tissues and species (Chapter 2). The enzyme from *Candida utilis* has been well studied (Itaya *et al.*, 1971). It has a molecular weight of 120,000, containing Fe^{3+}. Its optimal activity occurs at pH 8.5, and the K_m value toward urate is 6×10^{-6} *M*. Just like the enzymes purified from mammalian tissues and bacteria (Vogels and van der Drift, 1976), the *Candida* enzyme utilizes only oxygen as the electron acceptor, and has a low specific enzyme activity of 10.5 IU/mg protein. Although not as well studied, the enzymes purified or partially purified from other fungal species (Vogels and van der Drift, 1976), soybean nodules and radicles (Tajima and Yamamoto, 1975), and castor bean endosperm (Theimer and Beevers, 1971) generally have properties similar to those of the *Candida* enzyme.

D. D-Amino Acid Oxidase

$$RCHNH_2COOH + O_2 + H_2O \rightarrow RCOCOOH + NH_3 + H_2O_2 \quad (4.6)$$

This flavoenzyme in mammalian tissues has been studied quite extensively (Massey *et al.*, 1961; Curti *et al.*, 1973), but its presence in the peroxisomes or in tissue extracts of higher plants has not been documented. The mammalian enzyme has a molecular weight of 90,000, containing two FAD moieties. The maximal activity occurs at pH 8.5–9.0. The enzyme has a broad substrate specificity toward many D-isomers of amino acids. The enzyme is localized in the peroxisomes of some fungi, algae, and mammalian tissues (Section VII). Its specific activity in isolated rat liver or kidney peroxisomes is low in comparison with other enzymes. The physiological role of the enzyme in mammalian tissues as well as in fungi and algae is unknown.

III. UNSPECIALIZED PEROXISOMES

As mentioned in Chapter 2, peroxisomes known to contain only catalase and limited activities of H_2O_2-producing oxidases are present as minor cellular constituents in a variety of tissues. Apparently, they do not carry out an active metabolism in the sense of massive metabolite flux as do the leaf peroxisomes or glyoxysomes. Several possible functions of these peroxisomes have been suggested (Tolbert, 1971). The organelles catabolize metabolites like glycolate and urate that cannot be oxidized by the mitochondria. During peroxisomal oxidation, no useful energy in the form of ATP or reduced pyridine nucleotides is produced, and thus the peroxisomes may represent a means of disposing excess cellular energy. Because some of the H_2O_2-producing oxidases have a low affinity for oxygen, the peroxisomes may protect the cells from oxygen toxicity under high oxygen tension.

Although no massive metabolite flux appears to occur in these peroxisomes, the organelles may still perform unique and important functions that have yet to be discovered. Many cellular processes do not involve massive metabolite fluxes but nevertheless serve indispensable physiological functions, such as hormone production. The unspecialized peroxisomes might turn out to be important physiologically, but whether or not it is the case requires further research.

IV. GLYOXYSOMES IN GERMINATED OILSEEDS

A. Gluconeogenesis from Reserve Lipids

The storage tissues of oilseeds possess a gluconeogenic pathway for the conversion of stored lipid to sugar during postgerminative growth. The gluconeogenic pathway involves many enzymes and organelles. Among the storage tissues of various oilseeds, the pathway in the endosperm of castor bean seedlings has been the most completely studied. The relatively large castor bean endosperm is ideally suited for organelle isolation in that the soft tissue is composed mostly of one type of parenchyma cells containing numerous organelles. Furthermore, the pastids and other endomembrane systems are poorly developed such that they would not be available to contaminate the glyoxysomal fraction, as is known to be the case in some other oilseeds. The main function of the castor bean endosperm is the storage and mobilization of reserve lipids and proteins. After the mobilized reserves have been transported to the absorbing cotyledons and ultimately to the growing embryonic axis, the tissue senesces. This is in contrast to oil-storing cotyledons of some other seeds in which the cotyledons differentiate into photosynthetic organs as the reserves are mobilized. The ultimate fate of some of the reserves in these cotyledons may well be different than from that in the castor bean endosperm, but the main metabolic role of the glyoxysomes in processing fatty acid and acetate derived from the reserve lipids appears to be similar in the two types of tissues.

In this and the next few sections, the role of castor bean glyoxysomes in mobilizing and converting reserve lipid to carbohydrate (gluconeogenesis) will be emphasized. Appropriate comparisons with similar data on glyoxysomes isolated from other oil seeds are made. In Section IV,G, similarities and differences among oilseeds in processing the storage triacylglycerols will be summarized and discussed.

The gluconeogenic pathway in castor bean endosperm involves many enzymes in several subcellular compartments including lipid bodies (spherosomes), glyoxysomes, mitochondria, and the cytosol (Fig. 4.1). Briefly, the fatty acids liberated from triacylglycerols in the lipid bodies are oxidized through the glyoxysomal β-oxidation sequence producing acetyl-CoA, which is metabolized through the glyoxylate cycle, within the glyoxysomes. The end product succinate leaves the glyoxysomes, enters the mitochondria, and is oxidized to oxaloacetate through part of the Krebs cycle. In the cytosol, oxaloacetate is decarboxylated to form phosphoenolpyruvate which is converted to hexose via reversed glycolysis. In castor bean, there is essentially a quantitative conversion of triacyglycerol to carbohydrates on a gram-to-gram basis, suggesting a remarkably efficient gluconeogenic process. The succinate generated in the glyoxysomes probably is used exclusively for gluconeogenesis to form sucrose in castor bean endosperm, but is partly diverted to other metabolic processes in the cotyledons of many oilseeds. Nevertheless, glyoxysomal metabolism essentially is the same in all these tissues.

B. Hydrolysis of Triacylglycerols to Fatty Acids

The gluconeogenic pathway begins with lipid hydrolysis. Uncertainties exist as to which subcellular compartment contains the enzyme (lipase) for the first step in triacylglycerol hydrolysis and what is the initial substrate of glyoxysomal metabolism. The available evidence indicates that the mechanism is different among species.

In castor bean endosperm, an acid lipase is associated with the membrane of the storage lipid bodies (oleosomes, spherosomes). The activity of this lipase is already high in the dry seed and the enzyme can hydrolyze tri-, di-, and monoacylglycerols (Ory *et al.*, 1968). Under optimal *in vitro* conditions, the lipid bodies isolated from dry seeds can hydrolyze

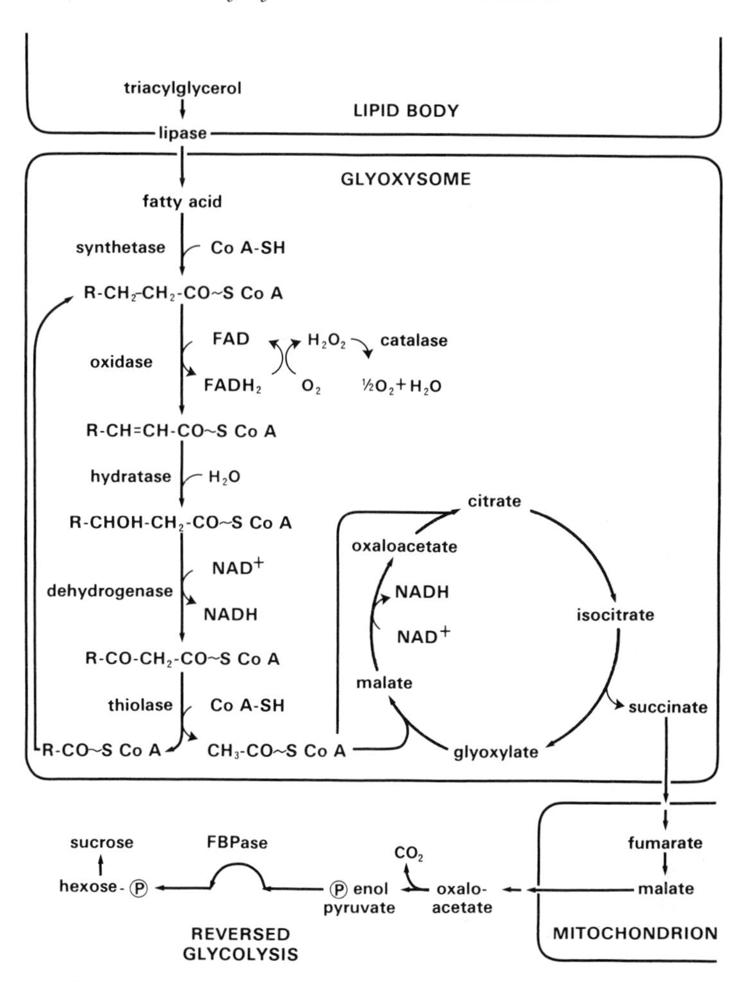

Fig. 4.1. Schematic representation of cellular compartments and enzymes involved in gluconeogenesis from storage lipid in the endosperm of castor bean seedlings. The five enzymes of the glyoxylate cycle are aconitase for citrate to isocitrate, isocitrate lyase for isocitrate to glyoxylate, malate synthase for glyoxylate to malate, malate dehydrogenase for malate to oxaloacetate, and citrate synthase for oxaloacetate to citrate conversions.

all of their triacylglycerols within ½ hr. The acid lipase activity decreases following germination, concomitant with the loss of storage triacylglycerols (Moreau *et al.*, 1980). On the other hand, the glyoxysomes possess an alkaline lipase (Muto and Beevers, 1974). This enzyme can hydrolyze trilinolein, dilinolein, monolinolein, and monopalmitin, but not di- or tripalmitin (Huang, 1983); the activity on trilinolein is about 10% that on monolinolein. Its activity on trilinolein is only a few percent of the lipid body acid lipase. It is possible that the storage triacylglycerol is hydrolyzed first by the lipid body acid lipase to one monoacylglycerol and two fatty acids which are then transported to the glyoxysomes. Subsequently, the glyoxysomal alkaline lipase would hydrolyze the monoacylglycerol to glycerol, which is released to the cytosol, and a third fatty acid, which is oxidized in the glyoxysomes. However, it is difficult to visualize how the monoacylglycerol can escape the catalytic activity of the lipid body acid lipase and move to the glyoxysomes, especially since the acid lipase shows a roughly equal *in vitro* activity on mono-, di-, and tripalmitins (Muto and Beevers, 1974). It is even more difficult to understand why there is a need for the alkaline lipase in the glyoxysomes when the hydrolysis could be easily accomplished by the lipid body acid lipase alone.

The pattern of acid and alkaline lipase in castor bean does not seem to be common among oilseeds. Acid lipase similar to the lipid body enzyme of castor bean is absent in many oilseeds examined (Akhtar *et al.*, 1975; Huang, 1983). Oilseeds in which the lipolysis has been examined in some detail can be divided roughly into two groups according to the pattern of lipolysis. One group has lipolysis activity associated with the lipid bodies in germinated but not ungerminated seeds, and the activity is highest at neutral (corn and rapeseed) or acidic (mustard and cotton) pHs. The other group includes seeds of soybean, peanut, cucumber, sunflower, and pine, which have no lipolytic activity at pH 5, 7, or 9 in the lipid bodies isolated from ungerminated or germinated seeds. Glyoxysomes isolated from soybean, peanut, and cucumber contain alkaline lipase activity, and the soybean enzyme can hydrolyze trilinolein, dilinolein, monolinolein, and monopalmitin. The glyoxysomal lipase activities in these seeds are quite low in comparison with the lipid body lipase activity in other seeds. The glyoxysomes from pine and sunflower seeds do not contain lipase activity. In pine seeds, soluble lipase activity has been reported (Ching, 1973).

In the cotyledons of rape, watermelon, and sunflower seeds, it was observed that the lipid bodies contained membrane appendices (Fig. 4.2) that were continuous with the lipid body membrane (Wanner and Theimer, 1978). Following germination, ribosomes appeared on the surface of the appendix. These ribosomes were postulated to synthesize the lipase which would then be associated with the appendix membrane. The appendix was proposed to be very fragile, and it was detached from the lipid body during homogenization, thus explaining the lack of lipase activity in isolated lipid bodies (Theimer and Rosnitschek, 1978). However, a subsequent study on rapeseed shows that more than 50% of the lipase activity was recovered in the lipid body fraction (Huang, 1983).

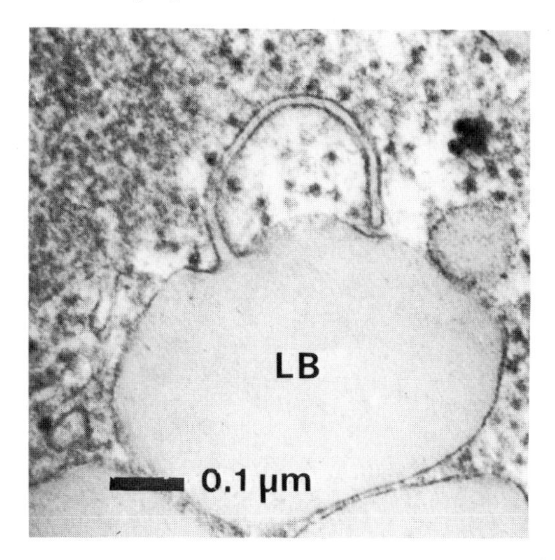

Fig. 4.2. Electron micrograph of lipid bodies (LB) (spherosomes) with a membraneous appendix in 3-day-old rapeseed seedlings. Note the continuation of lipid body membrane with the membrane of the appendix. Courtesy of G. Wanner.

Another report on mustard seed suggests that the appendix may have derived from lipid-depleted lipid-body sacules rather than from intact undigested lipid bodies (Bergfeld *et al.*, 1978).

The fatty acids of storage triacylglycerols from various oilseeds usually contain double and triple carbon bonds, and occasionally extra hydroxyl groups. These fatty acids may need to be modified prior to oxidation via the β-oxidation system. Information on the modification of fatty acids is meager. The modification may occur on the triacylglycerol directly in the lipid bodies before hydrolysis. More likely, the triacylglycerols are hydrolyzed to fatty acids, which are then modified in lipid bodies and/or glyoxysomes before or during an intermediate step of β-oxidation. In castor bean, the major storage fatty acid, ricinoleic acid, has one extra hydroxyl group at C-12 that is not conjugated with the double carbon bond at C-9 to C-10. An enzyme that converts the ricinolic acid to an intermediate that can be metabolized via β-oxidation was found in the castor bean glyoxysomes (Hutton and Stumpf, 1971). In this report, a balance sheet of the distribution of the total enzyme activity in various subcellular fractions was not presented, and thus further clarification for the subcellular localization of the enzyme is required. In cottonseed, the glyoxysomes contain a 3-*cis*-2-*trans*-enoyl-CoA isomerase, which isomerizes unsaturated fatty acids to a form that can be metabolized by subsequent β-oxidation (Miernyk and Trelease, 1981a). In jojoba seed, which stores wax esters instead of triacylglycerols, the modification of fatty alcohols (from the hydrolysis of wax ester) to fatty acids occurs in the wax bodies and not in the glyoxysomes (Moreau and Huang, 1979). However, in yeast grown on alkane, the modification of fatty alcohol (from alkane) to fatty acid occurs in the glyoxysome-like peroxisomes (Section VII,B).

C. Metabolic Reactions in Glyoxysomes

In castor bean, the fatty acids released from the storage triacylglycerols are metabolized exclusively in the glyoxysomes. They are first activated to fatty acyl-CoA by a fatty acyl-CoA synthetase in the presence of CoA, $MgCl_2$, and ATP (Cooper, 1971). This synthetase is specific for fatty acids of chain length greater than C_{10}. In addition to the above synthetase, there is an acetyl-CoA synthetase in the glyoxysomes, but its physiological function is unclear.

Fatty acyl-CoA is the substrate for the β-oxidation sequence consisting of four enzymes, two of which are in a complex. The reaction sequence can be assayed by fatty acyl-CoA-dependent NAD reduction, oxygen uptake, or acetyl-CoA formation (Fig. 4.1). The first reaction is catalyzed by fatty acyl-CoA oxidase that utilizes molecular oxygen as the direct electron acceptor to produce H_2O_2, which is apparently degraded by catalase within the organelles. A stoichiometry of ½ mole of O_2 uptake per mole fatty acyl-CoA oxidized is observed. The peroxisomal fatty acyl-CoA oxidase has been purified from rat liver and shown to be a flavoprotein with weakly bound FAD (Inestrosa *et al.*, 1980). It has not been purified from glyoxysomes, but indirect evidence also suggests that the castor bean enzyme contains a flavin prosthetic group for the transfer of electrons to O_2 (Cooper and Beevers, 1969b). The second and third enzymes (enoyl-CoA hydratase and β-hydroxyacyl-CoA dehydrogenase) have been purified to apparent homogeneity from cucumber cotyledons (Table 4.3) (Frevert and Kindl, 1980b). Both enzyme activities can be detected in the same protein of molecular weight 75,000; no subunit structure has been observed. Enoyl-CoA hydratase exerts higher activities toward substrates of shorter chain length (Miernyk and Trelease, 1981c). The fourth (last) enzyme of the β-oxidation sequence is thiolase (acetyl-CoA acetyltransferase). The enzyme purified from cucumber cotyledon has a molecular weight of 90,000 composed of two identical subunits (Frevert and Kindl, 1980a).

The fatty acyl-CoA synthetase and the β-oxidation enzymes are now considered to be located exclusively in the glyoxysomes of oilseeds. In earlier studies, β-oxidation activity was reported to be present in the glyoxysomes as well as in the mitochondria of castor bean endosperm (Hutton and Stumpf, 1969), pine megagametophytes (Ching, 1970), and corn scutella (Longo and Longo, 1975). In these studies, possible contamination of the mitochondria (1.19 g/cm^3) in sucrose gradients by broken glyoxysomes could not be ruled out, since the gradients were insufficiently fractioned. In more recent work, individual enzymes of the β-oxidation sequence were analyzed in numerous fractions from sucrose gradients (from pine megagametophyte, cotton cotyledon, and cucumber cotyledon), and the activities were found in the glyoxysomes, not in the mitochondria (Huang, 1975; Miernyk and Trelease, 1981a; Frevert and Kindl, 1980b). It is apparent, therefore, that the mitochondria in the storage tissues of oilseeds do not possess β-oxidation capacity and do not use acetate from fatty acids of storage triacylglycerols as a substrate for the Krebs cycle. It should be noted that in mammalian tissues, fatty acyl-CoA synthetase and the β-oxidation enzymes occur in both the mitochondria and peroxisomes (Tolbert, 1981). In the mitochondria, the first β-oxidation reaction is catalyzed by fatty acyl-CoA dehydrogenase that is coupled to the electron transport chain, whereas in the peroxisomes, the enzyme is an oxidase utilizing molecular oxygen as the direct electron acceptor (as is the glyoxysomal enzyme).

The product of β-oxidation, acetyl-CoA, is metabolized via the glyoxylate cycle. The enzymes of the glyoxylate cycle also are restricted to the glyoxysomes. Two of the enzymes (malate synthase and isocitrate lyase) are unique to the glyoxylate cycle, whereas the other three enzymes (malate dehydrogenase, aconitase, and citrate synthase) also are components of the Krebs cycle in the mitochondria. All the glyoxylate cycle enzymes except aconitase have been purified at least partially and their properties studied (Table 4.3). Aconitase is unstable during the preparation of glyoxysomes, but the loss in activity can be partially prevented by the use of sulfhydryl reagent (Cooper and Beevers, 1969a). Furthermore, the enzyme is easily solubilized during organelle preparation. The isozymic nature of the glyoxysomal malate dehydrogenase and citrate synthase (Table 4.4) has been reported (reviewed by Ting *et al.*, 1975; Kagawa and Gonzalez, 1981). The various isozymic forms of each enzyme within the same tissue can be separated by gel electrophoresis, gel filtration, or ion-exchange chromatography. The distinction among the isozymes lies in the electrophoretic mobility, heat stability, subunit size, immunological kinship, and kinetic properties. Both glyoxysomal malate dehydrogenase and citrate synthase have a low electrophoretic mobility at neutral pH on starch gel, and a great tendency to self-aggregate.

In essence, the glyoxylate cycle converts two molecules of acetate to

TABLE 4.4

Isozymes in Peroxisomes and Other Subcellular Compartments in Castor Bean Endosperm and Spinach Leaves

	Glyoxysomes	Leaf peroxisomes	Mitochondria	Plastids	Cytosol
Malate dehydrogenase	+	+	+		+
Aspartate–α-ketoglutarate Aminotransferase	+	+	+	+	+
Aconitase[a]	+		+		
Citrate synthase	+		+		

[a] The isozymic nature of aconitase is speculative. Malate dehydrogenase is NAD-specific. The plastids contain a NADP-specific malate dehydrogenase.

one molecule of succinate, the end product of the cycle (Fig. 4.1). In accordance with succinate being the end product, isolated glyoxysomes were shown to metabolize [5,6−^{14}C]isocitrate in the presence of acetyl-CoA to radioactive succinate (presumably 1,4−^{14}C-labeled) and not malate (Fig. 4.3).

Succinate leaves the glyoxysomes and is converted to malate or oxaloacetate by Krebs cycle enzymes in the mitochondria. It is more likely that malate and not oxaloacetate is the end product of this mitochondrial metabolism. Glyoxysomes alone convert [5,6−^{14}C]isocitrate to radioactive succinate only, whereas addition of mitochondria to the glyoxysomes results in further metabolism of succinate to malate (Fig. 4.3). The subsequent reactions for the conversion of oxaloacetate to sucrose occur in the cytosol. Oxaloacetate is decarboxylated to phosphoenolpyruvate catalyzed by phosphoenolpyruvate carboxykinase. Phosphoenolpyruvate is metabolized to hexose phosphate through reversed glycolysis, requiring fructose 1,6-bisphosphatase to overcome the nonreversible reaction catalyzed by fructose phosphate kinase (Fig. 4.1) (Thomas and ap Rees, 1972; Youle and Huang, 1976). Sucrose is the major product of the gluconeogenic pathway.

D. Oxidation of NADH Generated in Glyoxysomes by Electron Shuttle

Two reactions in the glyoxysomes reduce NAD to NADH. For each sequence of the β-oxidation, one NADH is generated, and for each turn of the glyoxylate cycle, another NADH is produced (Fig. 4.1). The glyoxysomes do not have the capacity to reoxidize NADH (Lord and Beevers,

1972), and they contain only a catalytic amount of NAD/NADH (0.2–0.6 nmoles/mg protein) (Mettler and Beevers, 1980; Hicks and Donaldson, 1982). The limited activity of NADH-cytochrome reductase is not sufficient to account for the rate of NADH re-oxidation (Hicks and Donaldson, 1982). Apparently, the NADH is reoxidized rapidly outside the glyoxysomes, presumably in the mitochondria.

A highly active aspartate aminotransferase is present in the glyoxysomes of castor bean (Cooper and Beevers, 1969a) and other oilseeds (Gerhardt, 1978). The enzyme does not participate directly in the gluconeogenic pathway. It is highly specific for α-ketoglutarate as the amino group acceptor, and it occurs as an isozyme form that is distinct from those of other subcellular compartments (Table 4.4). This aminotransferase, together with malate dehydrogenase, has been suggested to form an electron shuttle system for the reoxidation of NADH generated in the glyoxysomes. The system, utilizing malate and aspartate as the vehicle, occurs in the mitochondria and also has been proposed to occur in the leaf peroxisomes (see Section V,C and Fig. 4.6). Another electron shuttle system coupled with an incomplete glyoxylate cycle (Fig. 4.4) recently has been proposed (Mettler and Beevers, 1980). The evidence for this proposed shuttle is based heavily on the equilibrium of the malate dehydrogenase reaction in the glyoxysomes strongly favoring malate formation. In the scheme, Krebs cycle enzymes in the mitochondria are not involved directly in the flow of carbon in gluconeogenesis; rather, they participate in the electron transport system. A potential weakness in the proposal is the fact that the malate dehydrogenase in the Krebs cycle also catalyzes a reaction with an equilibrium that greatly favors malate formation; yet, the reaction does go to oxaloacetate formation during the normal operation of the Krebs cycle. Further experimentation is required to test the validity of the scheme,

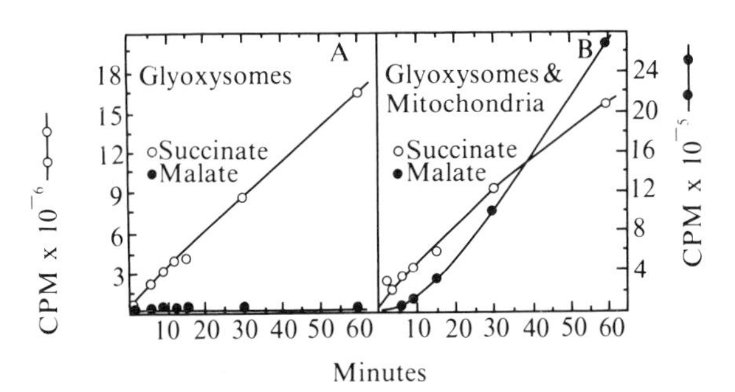

Fig. 4.3. Conversion of [5,6−^{14}C]isocitrate to [^{14}C]succinate and [^{14}C]malate by organelles isolated from castor bean endosperm. The reaction mixture also contained acetyl-CoA. Redrawn from Cooper and Beevers (1969a).

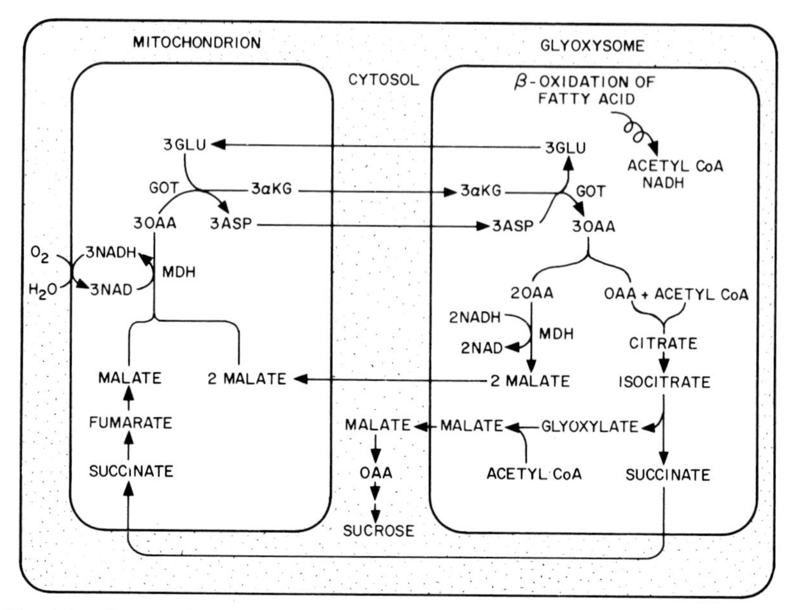

Fig. 4.4. Proposed malate–aspartate shuttle for the transfer of reducing equivalents between glyoxysomes and mitochondria providing for *in situ* oxidation of NADH produced by glyoxysomal β-oxidation. GOT stands for aspartate–α-ketoglutarate aminotransferase, and MDH represents malate dehydrogenase. From Mettler and Beevers (1980).

especially the detection of a transport system on the glyoxysomal membrane and the reconstruction of the electron shuttle system *in vitro* using intact glyoxysomes.

E. Characteristic Peroxisomal Enzymes in Glyoxysomes

The glyoxysomal catalase probably serves a specific function in degrading the H_2O_2 generated by fatty acyl-CoA oxidase during the β-oxidation of fatty acids (Cooper and Beevers, 1969b). In addition to catalase, glyoxysomes contain other characteristic peroxisomal enzymes including urate oxidase, glycolate oxidase, allantoinase (an enzyme in urate degradation), and hydroxypyruvate reductase (an enzyme important in the photorespiratory pathway; see Section V,B). Except for catalase, these peroxisomal enzymes are present in low specific activities in isolated glyoxysomes (Tables 4.1 and 4.2). Their low activities contrast with the relatively high activities of urate oxidase in mammalian liver peroxisomes and root nodule peroxisomes, or glycolate oxidase and hydroxypyruvate reductase in leaf peroxisomes; the liver, root nodule, and leaf peroxisomes are specialized to perform specific physiological functions requiring the high enzyme activities. The existence of these enzymes in the glyoxysomes reinforces the concept that glyoxysomes, although having a specialized physiological function, are a type of peroxisome.

F. Enzymes Associated with the Glyoxysomal Membrane

Of the various peroxisomes in plants, animals, and microorganisms, only glyoxysomes from oilseeds have been shown to have highly active enzymes associated with the surrounding membranes. The most complete work has been done using glyoxysomes from castor bean endosperm (Huang and Beevers, 1973; Bieglmayer *et al.*, 1973). Subsequent comparative studies using glyoxysomes from other oilseeds have revealed similar findings but with some differences (Huang, 1975). Most castor bean glyoxysomes at the peak stage of organelle development following germination do not contain crystalline cores (Chapter 2), and thus there are only two major suborganelle compartments, the membrane and the matrix. Glyoxysomes, normally isolated in 54% sucrose, can be broken osmotically by simple dilution with water, and the solubilized matrix enzymes are readily separated in the supernatant from the sedimented membrane by centrifugation (Table 4.5). The membranes in the pellet appear as glyoxysomal ghosts having approximately the original size of the organelles (Fig. 3.8). Many enzymes are associated with the membrane with varying degrees of tightness. Different solubilization agents, such as detergents or salts, at the appropriate concentration preferentially remove most of the membrane-associated enzymes. In castor bean glyoxysomes, alkaline lipase is bound to the membrane most tightly; it resists solubilization by high concentration of salt (1.0 *M*) and is inactivated by detergents (Muto and Beevers, 1974). Malate synthase and citrate synthase are associated with the membrane and can be removed with 0.15 *M* KCl (Fig. 4.5). Other enzymes, including malate dehydrogenase and several of the β-oxidation enzymes, are only loosely associated with the membrane and can be removed easily with salt concentration as low as 0.05 *M*. In essence, many enzymes of the gluconeogenic pathway in the glyoxysomes are membrane associated. In oilseeds other than castor bean, the same conclusion can be made, even though the apparent degrees of association of enzymes with the membrane vary (Huang, 1975; Longo *et al.*, 1975; Bortman *et al.*, 1981). In castor bean and a few other oilseeds, KCl at 0.2 *M* and Triton X-100 at 0.05% generally increase the *in vitro* activities of malate synthase, citrate synthase and β-hydroxyacyl-CoA dehydrogenase (which are membrane associated) in isolated glyoxysomes.

In all oilseeds examined, isocitrate lyase and thiolase, two enzymes of

TABLE 4.5

Percent Solubilization of Peroxisomal Enzyme Activities after Osmotic Breakage of Peroxisomes Isolated from Various Tissues[a]

	Potato tuber	Castor bean endosperm	Spinach leaves	Soybean nodule	Rat liver
Catalase	70	70	94	100	80
Glycolate oxidase	96	89	91	—[b]	63
Hydroxypryruvate reductase	—	93	98	—	—
Urate oxidase	88	94	—	100	0
Aspartate-α-ketoglutarate aminotransferase	—	78	86	—	—
Malate dehydrogenase	—	35	91	—	—
Isocitrate lyase	—	98	—	—	—
Thiolase	—	90	—	—	—
Enoyl-CoA hydratase	—	41	—	—	—
β-Hydroxyacyl-CoA dehydrogenase	—	43	—	—	—
Citrate synthase	—	11	—	—	—
Malate synthase	—	6	—	—	—
Protein	72	60	—	—	—

[a] The peroxisomal preparation in 50–54% sucrose was mixed with two volumes of dilute buffer, and centrifuged at 145,000 *g* for 60 min (soybean nodules) or 40,000 *g* for 30 min (the rest). Results are expressed as the amount in the supernatant divided by the sum of the amounts in the supernatant and in the pellet (Huang and Beevers, 1973; Hanks *et al.*, 1981).

[b] Activities not tested.

the gluconeogenic pathway, are solubilized readily from the glyoxysomes during osmotic shock (Table 4.5). It should be noted that an enzyme that can be solubilized easily by osmotic shock is not necessarily a matrix enzyme; the enzyme simply may be loosely associated with the outer surface of the membrane. No evidence is available to make this distinction. Catalase, urate oxidase, and glycolate oxidase, the three enzymes common to most peroxisomes, are easily solubilized from glyoxysomes.

It thus seems that many of the reactions of fatty acid β-oxidation and of the glyoxylate cycle are carried out in association with the membrane. However, the exact arrangement of these enzymes within the membrane, the rationale for such an arrangement, or the permeability of the membrane to various metabolites and cofactors are unknown.

Besides alkaline lipase, several enzymes are also tightly bound to the membrane of the glyoxysomes. These enzymes are present in low activities, and include cytochrome b_5, NADH-cytochrome reductase,

NADH-ferricyanide reductase (Hicks and Donaldson, 1982), cinnamic acid 4-hydroxylase, and *p*-chloro-*N*-methylaniline-*N*-demethylase (Young and Beevers, 1976). Rather than having important metabolic roles, these enzymes have been looked on as remnants of the endoplasmic reticulum membranes in biogenesis (see Chapter 5, Section III,A,2).

G. Processing of Storage Triacylglycerols in Oilseeds Other than Castor Bean

The pathway of gluconeogenesis and the operation of glyoxysomes elucidated for the castor bean endosperm generally are assumed to be similar for oil-storage tissues in other seeds. The available data support this assumption, with only minor exceptions. However, the overall metabolic fate of carbon originated from storage triacylglycerol after having processed in the glyoxysomes may vary depending on the function of the tissue.

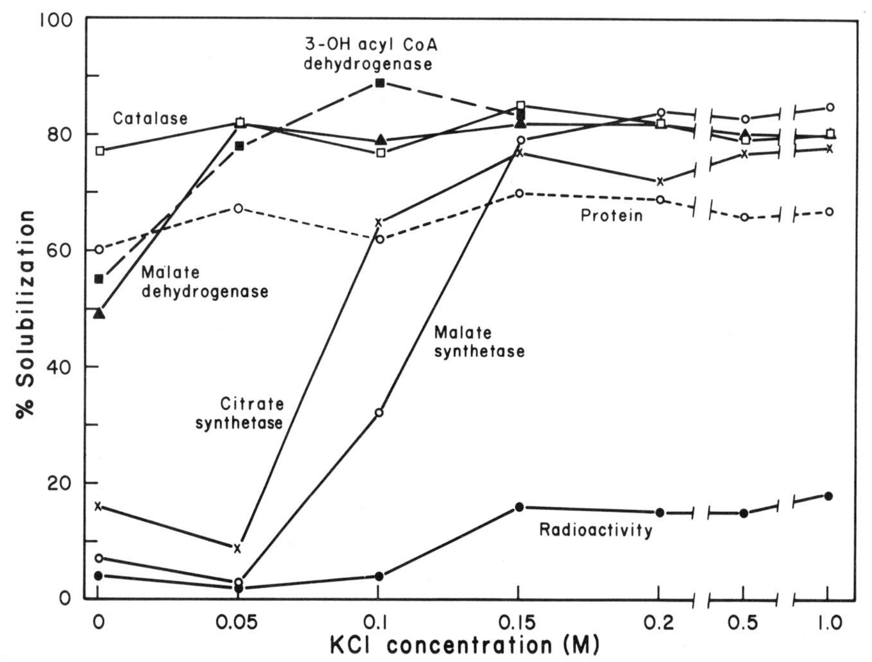

Fig. 4.5. Solubilization of enzymes when isolated castor bean glyoxysomal fraction in 54% sucrose solution was mixed with two volumes of dilute buffer containing various concentrations of KCl. The concentration of KCl shown is the final concentration after mixing. The radioactivity represents the membrane in which the lecithin had been labeled by previous treatment of the tissue with [^{14}C]choline. From Huang and Beevers (1973).

In castor bean endosperm, whose sole function is storage and mobilization of food reserves, storage lipid is converted to sucrose through the gluconeogenesis pathway (Fig. 4.1), and then translocated to the growing axis. According to the known gluconeogenic pathway (Fig. 4.1), 25% of the fatty acid carbon is released as CO_2 during the conversion of oxaloacetate to phosphoenolpyruvate, and the other 75% reappears in sucrose. This 75% conversion of lipid-to-carbohydrate carbon is in agreement with the observed respiratory quotient of 0.36 in endosperm tissue (Beevers, 1961), assuming that a C_{16} saturated fatty acid is the substrate (castor bean contains mostly ricinoleic acid, $C_{18}H_{34}O_3$):

$$C_{16}H_{32}O_2 + 11\ 0_2 \rightarrow C_{12}H_{22}O_{11} + 4\ CO_2 + 5\ H_2O \qquad (4.7)$$

The respiratory quotient value suggests that the CO_2 released in the conversion of oxaloacetate to phosphoenolpyruvate is not refixed, unless the CO_2 released is derived from substrates other than fatty acids. A 100% weight conversion of lipid to carbohydrate (gram-to-gram) can be obtained (Beevers, 1980) and is due to the extra oxygen atoms in sugar molecules making up for the carbon loss from the fatty acid. Additional enrichment of the recovery comes from the conversion of glycerol (Beevers, 1956) and amino acids of the storage protein (Stewart and Beevers, 1967).

This conversion of storage triacylglycerols solely to sugars in castor bean endosperm does not occur in the storage tissues of some other oilseeds. In cotyledons of oilseeds such as sunflower, cotton, and the cucurbits (watermelon, cucumber, marrow, etc.), the storage cotyledon differentiate into photosynthetic organs in response to light as the food reserves are depleted. In these cases, the reserves are used to support growth of the embryo axis as well as formation of the photosynthetic machinery. The conversion of triacylglycerol to carbohydrate and other substances in excised marrow cotyledons has been examined (Thomas and ap Rees, 1972). Only a small portion of the degraded triacylglycerol (15%) was recovered in soluble carbohydrate (mostly sucrose and stachyose); the remainder was in lipids, amino acids, nucleotides, proteins, and released CO_2. The contribution to cell wall components and starch was not examined. Unfortunately, this study with excised marrow cotyledons did not permit an assessment of the proportion of lipid catabolites used for maintenance and differentiation of the cotyledon and growth of the axis. Recent studies with [2−^{14}C]acetate supplied to intact cotton seedlings (Doman and Trelease, 1982) revealed that some 30–35% of the ^{14}C was chased into the radicles, suggesting that a sizable percentage of the triacylglycerol was converted to substances that remained in the cotyledons. Nevertheless, as judged from the time course and specific radioactive labeling pattern, the acetate derived from storage triacylglycerols in cotton cotyledon likely is processed exclusively to succinate by the glyoxylate cycle in the glyoxysomes and is then converted to sucrose and other metabolites; these metabolites are used either directly in the cotyledons or exported to the growing embryonic axis. In corn scutella after germination, a quantitative conversion of lipid to carbohydrate also does not occur, although sucrose is a main product (Oaks and Beevers, 1964). Similar studies have not been done with those triacylglycerol-storing tissues that do not become photosynthetic, such as the megagametophyte in pine, perisperm in yucca, and cotyledons in peanut. These tissues probably are more like the castor bean endosperm in carrying out the major function of mobilizing food reserve for transport to the growing axes.

Although there are substantial variations among various oilseeds in the utilization of the gluconeogenic products from triacylglycerols, the glyoxysomes occurring in these oilseeds are remarkably similar. A comparative study of the enzymes in glyoxysomes isolated from various tissues of oilseeds shows that the specific activities of individual enzymes are strikingly similar (Table 4.6). In one or two oilseeds, the specific activities of the various enzymes are slightly lower, possibly due to contamination of the isolated glyoxysomes by other organelles, but the relative activities of the enzymes still are similar to those of other oilseeds. The few exceptions are the activities of alkaline lipase (its function and native substrate are still unknown) and cotton malate dehydrogenase, and no explanation has been given. Glyoxysomes isolated from various oilseeds also show qualitative similarities, though with quantitative differences, in the tightness of enzyme association with the membrane and the activation of these enzymes by salt and detergent (Huang, 1975; Longo and Longo, 1975; Bortman *et al.*, 1981). Apparently, glyoxysomes in various oilseeds operate in a similar fashion.

Peroxisomes containing β-oxidation and glyoxylate cycle enzymes are also present in late-stage maturing embryos of cotton (Choinski and Trelease, 1978; Miernyk and Trelease, 1981a) and cucumber (Köller *et al.*, 1979). These peroxisomes in cucumber have been looked on as "incomplete glyoxysomes," being intermediates of complete glyoxysomes by accumulating enzymes sequentially following germination (Köller *et al.*, 1979). Accordingly, they are considered metabolically nonfunctional during cucumber seed maturation. In cotton embryo, the peroxisomes are postulated to function in the synthesis of citrate, which accumulates during the late stage of maturation (Miernyk and Trelease, 1981b). The cotton organelles possessing malate synthase and citrate synthase have the capacity to synthesize citrate from glyoxylate or malate, as demon-

TABLE 4.6

Enzyme Activities in Glyoxysomes Isolated from Various Oilseeds[a]

Enzymes	Castor bean endosperm	Watermelon cotyledon	Peanut cotyledon	Cucumber cotyledon	Pine megagametophyte	Sunflower cotyledon	Jojoba cotyledon	Cotton cotyledon
Catalase	13,000,000	9,900,000	11,000,000	11,000,000	9,500,000	7,400,000	9,100,000	8,200,000
Malate dehydrogenase	38,000	55,000	34,000	80,000	60,000	16,000	33,000	3,800
β-Hydroxyacyl-CoA dehydrogenase	6,500	7,100	6,200	5,600	6,400	—	4,700	1,200
Malate synthase	2,100	1,100	1,300	1,600	2,100	1,560	1,600	742
Isocitrate lyase	930	520	300	330	480	820	1,100	120
Citrate synthase	840	700	330	800	350	426	—	239
Glycolate oxidase	95	70	58	89	—	327	—	—
Alkaline lipase	19	4	7	2	8	—	0	—

[a] Data are obtained from Schnarrenberger *et al.* (1971), Huang (1975), and Bortman *et al.* (1981).
[b] Activities not tested.

strated by *in vitro* and *in vivo* labeling experiments. However, why the mitochondrial citrate synthase was not functional in citrate synthesis and what the purpose of citrate accumulation is are not clear.

Gluconeogenesis from storage lipid also appears to occur in fern spores following germination. In the spores of *Onoclea sensibilis, Dryopteris filixmas,* and *Anemia phyllitidis,* peroxisomes are observed *in situ* adjacent to lipid bodies, and there is a concomitant depletion of storage lipid and increase in isocitrate lyase and malate synthase activities (DeMaggio *et al.*, 1980; Gemmrich, 1981). Presumably, the peroxisomes are similar to the glyoxysomes in oilseeds, but the organelles have not been isolated nor their enzyme content characterized. Whether sugar is a major product of the degraded lipid is not known.

V. LEAF PEROXISOMES

A. Photorespiration and Leaf Peroxisomes

The major physiological process that the leaf peroxisomes participate in is photorespiration. Photorespiration is the light-dependent O_2 uptake and/or CO_2 evolution. It occurs by a metabolic pathway that differs from that of dark respiration. The process proceeds at a faster rate under high light intensity, elevated temperature, and high O_2/CO_2 ratio. It is generally thought that photorespiration is a wasteful metabolic process due to the undesirable production of photorespiratory substrates. Newly formed photosynthates go through oxidative reactions to release the fixed CO_2. Photorespiration reduces the rate of CO_2 fixation by 10–50% in some plant species. These species utilize the Calvin cycle to fix CO_2 initially into C_3 acids, and are termed C_3 species. In other species that carry out photosynthetic CO_2 fixation efficiently, internal mechanisms are present to drastically reduce or eliminate photorespiration. In these efficient species, CO_2 is fixed initially into a C_4 acid prior to its entry to the normal Calvin cycle; they are termed C_4 species. In this section and Sections V,B and V,C, the mechanism of photorespiration in C_3 species will be described with emphasis on the role of leaf peroxisomes. In Section V,D, photorespiration and peroxisomes in the leaves of C_4 species will be discussed.

The initial substrate for photorespiration, phosphoglycolate, is produced in the chloroplasts (Fig. 4.6). The most popular hypothesis is that phosphoglycolate is generated by the oxygenase activity of ribulose 1,5-bisphosphate carboxylase/oxygenase (reviewed by Jensen and Bahr, 1977). The enzyme can catalyze the condensation of ribulose 1,5-

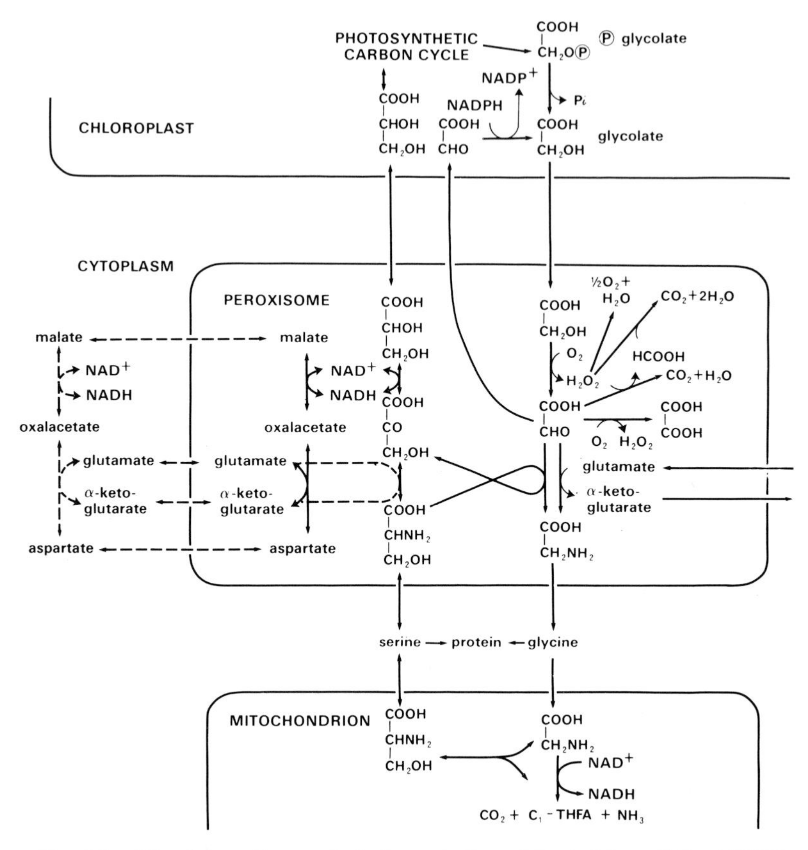

Fig. 4.6. Glycolate pathway of photorespiration. Adopted from Tolbert (1980).

bisphosphate with either CO_2 to form two molecules of 3-phosphoglyceric acid, or O_2 to form one molecule of 3-phosphoglyceric acid and one molecule of 2-phosphoglycolic acid. The amount of 2-phosphoglycolic acid produced *in vivo* depends partly on the relative concentration of O_2 and CO_2 in the vicinity of the enzyme. The 2-phosphoglycolate produced is hydrolyzed to glycolate by a specific phosphoglycolate phosphatase in the chloroplasts (Tolbert, 1980). Two other hypotheses of glycolate production were proposed but have received less support. One of them is the oxidation of the thiamine pyrophosphate–C_2 complex of transketolase in the Calvin cycle to produce a molecule of glycolate directly (Gibbs, 1969). The other hypothesis is the reductive condensation of two CO_2 to form glycolate (Zelitch, 1971).

Glycolate leaves the chloroplasts and enters the leaf peroxisomes (Fig. 4.6). The leaf peroxisomes contain highly active glycolate oxidase, which catalyzes the oxidation of glycolate in the presence of O_2 to produce H_2O_2 and glyoxylate. Under adequate nitrogen supply, most of the glyoxylate undergoes the glycolate pathway (Fig. 4.6). Glyoxylate is converted to glycine by a transamination reaction in the peroxisomes. Glycine leaves the peroxisomes and enters the mitochondria where two glycine molecules are condensed to form one serine molecule, one NH_3, and one CO_2. This is the CO_2 generated in photorespiration. Serine then returns to the peroxisomes where it is converted to hydroxypyruvate by transamination. Hydroxypyruvate is reduced by NADH to glycerate, which then departs from the peroxisomes and enters the chloroplasts, where it is converted eventually to sugars.

There were doubts on the glycolate pathway being the major route of further glyoxylate metabolism. It was suggested that glyoxylate goes back to the chloroplasts where it is oxidized to CO_2 (Zelitch, 1972) or reduced to glycolate in the presence of a relatively active and specific NADPH-glyoxylate reductase (Tolbert, 1980). The latter activity together with glycolate oxidase activity would constitute a glycolate-glyoxylate shuttle, serving the purpose of disposing excessive reducing equivalent generated in the chloroplasts during photosynthesis. Another suggestion was that glyoxylate is oxidized non-enzymatically by H_2O_2, which is present due to an incomplete elimination by catalase in the peroxisomes (Grodzinski and Butt, 1977). The products are CO_2 and formate (Fig. 4.7). Formate can be oxidized to CO_2 in the peroxisomes by the peroxidative activity of catalase in the presence of H_2O_2, or in the

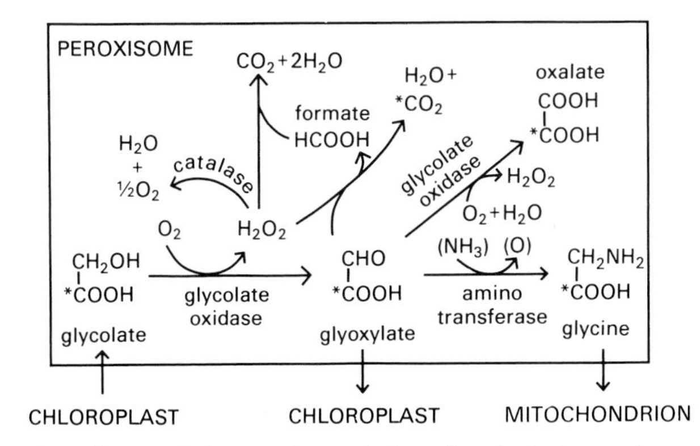

Fig. 4.7. Possible metabolic reactions of glyoxylate in leaf peroxisomes. Glyoxylate derived from glycolate can be converted to various metabolites in the peroxisomes, but it can also be metabolized in the chloroplasts. The oxidation of glyoxylate to formate and CO_2 occurs nonenzymatically. The oxidation of formate to CO_2 can occur either in the peroxisomes by the peroxidative activity of catalase in the presence of H_2O_2, or in the mitochondria by NAD-formate dehydrogenase. From Chang and Huang (1981).

mitochondria by NAD-formate dehydrogenase, or outside the peroxisomes by unknown reactions. In these suggestions, CO_2 is released directly from glyoxylate or formate as the photorespired CO_2. Another possible drainage of glyoxylate is its oxidation to oxalate by glycolate oxidase (see Section V.B.1) in the peroxisomes (Fig. 4.7). Spinach leaf discs have the ability to convert supplied radioactive glycolate or glyoxylate to oxalate (Chang and Beevers, 1968). After its formation in green leaves, oxalate is not degraded and can accumulate to a high amount, often in crystalline form (Zindler-Frank, 1976).

Recent studies strongly indicate that indeed the glycolate pathway is the major route of further glyoxylate metabolism in the presence of sufficient nitrogen supply. The evidence includes the kinetics of glycolate metabolism in isolated leaf peroxisomes (Chang and Huang, 1981; Walter, 1982), the characterization of *Arabidopsis* mutants deficient in photorespiratory enzymes (Somerville and Ogren, 1982), and the photorespiratory carbon flow in leaf discs and cells in the presence and absence of amino acids (Somerville and Ogren, 1982; Oliver, 1981). The drainage of glyoxylate to other metabolic reactions should be quite minor and have little impact on the photorespiratory carbon flow. Whether these reactions have more physiological significance on other aspects of leaf metabolism remains to be seen.

Although photorespiration generally is considered a wasteful process, useful physiological roles have been proposed (Tolbert, 1980). In the glycolate pathway (Fig. 4.6), three of the four carbons from two glycolate molecules can be rescued into one glycerate molecule. Also, the leaf peroxisomes decompose glycolate, which, at high concentration, is toxic to the cells. Furthermore, under high O_2 tension due to active photosynthesis, the glycolate pathway can dispose of the excess light energy and oxygen so that destructive photooxidation would not occur.

Aside from metabolizing glycolate, the leaf peroxisomes and the glycolate pathway also are important in the synthesis of some amino acids. In the glycolate pathway, glycine and serine formed can be used for other metabolic reactions (Fig. 4.6). In addition, during glycine oxidation in the mitochondria, the C_1 fragment in the methylene tetrahydrofolate complex is the precursor of all essential metabolic reactions requiring C_1 group in the cells. Since the conversion between serine and glycerate is reversible, glycerate can enter the peroxisomes and be converted to serine, and subsequently to methylene tetrahydrofolic acid and glycine. Such a reaction can occur in the absence of light (e.g., at night) as long as glycerate is available (e.g., from starch degradation).

B. Enzymes in Leaf Peroxisomes

In the study of glyoxysomes, the castor bean organelles have been investigated most extensively. The research on leaf peroxisomes also has been concentrated on one main species, spinach. Although leaf peroxisomes have been isolated from leaves and green cotyledons of many other species (Chapter 2), these studies have been limited in scope. Our current knowledge on the operation of the leaf peroxisomes is based heavily on the studies with spinach and it is generally assumed that the information is valid on other C_3 species as well. However, the fact that it is an assumption should be noted.

When leaf peroxisomes isolated from spinach were subjected to the same procedure used to study the suborganelle localization of glyoxysomal enzymes, most activities of the enzymes examined were readily solubilized (Table 4.5). Apparently, most of the enzyme activities are not associated with the membrane or the crystalloids (Huang and Beevers, 1973). Cytochemical localization of glycolate oxidase in several photosynthetic tissues (Fig. 2.7) also indicates that glycolate oxidase is in the matrix and not in the crystalloids (Thomas and Trelease, 1981). Diaminobenzidine cytochemistry shows that catalase is present in both the crystalloids and matrix (Frederick and Newcomb, 1969a). The relative amount of catalase activity in the crystalloids and matrix within a crystalloid-containing peroxisome, or within the whole peroxisome population, is unknown.

1. Conversion of Glycolate to Glycine

Glycolate oxidase and glyoxylate aminotransferase are the two enzymes involved in the conversion of glycolate to glycine. The properties of glycolate oxidase are described in Section II. Although glycolate oxidase also is present in peroxisomes of nongreen tissues, its specific activity in the leaf peroxisomes is more than 10 times higher than that in isolated peroxisomes of other plant and animal tissues (Table 4.1). Such a high activity signifies the importance of this enzyme in the leaf peroxisomes. The enzyme activity per gram fresh weight of leaves is high enough to account for all the metabolism of glycolate during active photorespiration (Zelitch, 1971). The reaction is irreversible under physiological conditions and the H_2O_2 produced is decomposed by catalase. The enzyme also can catalyze the oxidation of glyoxylate (the reaction product from glycolate) to oxalate, since glyoxylate occurs as a hydrated form in solution and thus appears as an α-hydroxyacid:

$$\underset{\text{glycolate}}{\begin{array}{c}CH_2OH\\|\\COOH\end{array}} + O_2 \longrightarrow \underset{\text{glyoxylate}}{\begin{array}{c}CHO\\|\\COOH\end{array}} + H_2O_2 \qquad (4.8)$$

$$\underset{\text{glyoxylate (hydrated form)}}{\begin{array}{c}CH(OH)_2\\|\\COOH\end{array}} + O_2 \longrightarrow \underset{\text{oxalate}}{\begin{array}{c}COOH\\|\\COOH\end{array}} + H_2O_2 \qquad (4.9)$$

This reaction probably does not occur readily due to the low affinity of the enzyme for glyoxylate (K_m value of 5.4×10^{-2} *M*, compared with 2×10^{-4} *M* for glycolate) (Richardson and Tolbert, 1961). The quantity of oxalate produced by glycolate oxidase depends on the amount of glyoxylate present in the peroxisomes; this amount in turn would depend on the extent of glyoxylate utilization (Fig. 4.7). An important control point would be the availability of glutamate or another amino group donor in the subsequent glyoxylate transamination (Fig. 4.6). The extent of oxalate synthesized by glycolate oxidase *in vivo* under physiological conditions remains to be seen. Whether a nitrogen deficiency enhances the synthesis and accumulation of oxalate in the leaves of spinach and other species also is unknown. If the flow of glyoxylate to oxalate is minimal, its contribution toward oxalate accumulation may still be significant over a long period of time. It should be noted that oxalate being derived from other metabolites also has been reported (Chang and Beevers, 1968; Loweus, 1980), and that leaves of many C_4 species that do not carry out active glycolate metabolism nevertheless accumulate oxalate (Zindler-Frank, 1976).

Leaf peroxisomes contain aminotransferases that are relatively specific for glyoxylate as the amino group acceptor. Two such enzymes have been separated from each other (Rehfeld and Tolbert, 1972), one being glutamate-glyoxylate aminotransferase and the other serine-glyoxylate aminotransferase. Since they are relatively, though not strictly, specific for glyoxylate, and are localized only in the peroxisomes, their assignment to the glyoxylate transamination is warranted (Tolbert, 1980). The specific activity of each enzyme in isolated peroxisomes in only about 10–20% of that of glycolate oxidase (Table 4.2). Concerns have been expressed on whether this transamination activity is sufficient to account for the rate of glycolate metabolism during photorespiration (Zelitch, 1971). If serine-glyoxylate aminotransferase is operative in the glycolate pathway as proposed, the reaction would be coupled to the conversion of serine to hydroxypyruvate (Fig. 4.7). However, since two glyoxylate molecules generate only one serine molecule, an extra amino group donor and aminotransferase are required, presumably glutamate and glutamate-glyoxylate aminotransferase.

Under the *in vitro* assay conditions, serine-glyoxylate aminotransferase and glutamate-glyoxylate aminotransferase reactions are not reversible. Thus, the conversion of glycolate to glycine is unidirectional. During steady-state condition of photosynthesis in green leaves, the pool of glycine is higher than that of glycolate (Tolbert, 1980).

2. Interconversion between Serine and Glycerate

The interconversion between serine and glycerate involves two reversible enzymatic reactions (Fig. 4.6). In the conversion of serine to hydroxypyruvate, several peroxisomal aminotransferases can be the catalyst. The glyoxylate-serine aminotransferase (see preceding two paragraphs) would be such an enzyme. Since this enzymatic reaction is irreversible in the formation of hydroxypyruvate and glycine, it would only operate in the direction of hydroxypyruvate formation in the proposed glycolate pathway. In the reverse reaction of serine formation, other aminotransferases are required. Peroxisomal aminotransferase utilizing glutamate or alanine as the amino group donor are present and can catalyze the reaction, but they are not specific for hydroxypyruvate (Rehfeld and Tolbert, 1972). More information is required on the enzymatic conversion between serine and hydroxypyruvate.

The reversible conversion between hydroxypyruvate and glycerate is catalyzed by NAD-linked hydroxypyruvate reductase (formerly known as glycerate dehydrogenase or NAD-glyoxylate reductase). The enzyme activity in isolated peroxisomes is about five times higher than that of glycolate oxidase (Tables 4.1 and 4.2). The enzyme from spinach leaves has been purified and well characterized (Kohn and Warren, 1970). It prefers NADH to NADPH as the electron donor. It can act on the substrate analog glyoxylate, and the apparent K_m value for hydroxypyruvate is 1.2×10^{-4} *M* and for glyoxylate, 5×10^{-2} *M*. With such a high K_m value for glyoxylate, the enzyme *in vivo* probably acts only on hydroxypyruvate to give rise to glycerate. Presumably, the conversion of glyoxylate to glycolate does not occur in the leaf peroxisomes, but does so in the chloroplasts where an NADP-glyoxylate reductase is present (K_m for glyoxylate, 1.3×10^{-4} *M*).

The combination of the above two reversible enzymatic reactions allows a bidirectional flow between serine and glycerate. The forward reaction toward glycerate formation is operative in the proposed glycolate pathway in photorespiration, whereas the backward reaction toward serine formation is functional in the dark for the generation of serine, glycine, and methylene tetrahydrofolic acid.

C. Electron Shuttle Involving Malate Dehydrogenase and Aspartate-α-ketoglutarate Aminotransferase

Information on the membrane permeability of the leaf peroxisomes (or any other peroxisomes) is still lacking, mainly due to the failure of obtaining pure and intact peroxisomes in dilute sucrose solution (see Chapter 3). The general assumption that the peroxisomal membrane is permeable to small metabolites such as glycolate, glycine, glycerate, and serine is purely speculative. The suggestion that the peroxisomal membrane is impermeable to NAD/NADH is based on our knowledge of the membrane permeability of chloroplasts and mitochondria. The only data available are that when intact peroxisomes (together with other organelles in a rate-sedimented pellet) were assayed for hydroxypyruvate reductase activity in an assay medium with or without osmotic fortification, broken organelles yielded some 10-fold higher activity of hydroxypyruvate-dependent NADH oxidation than intact organelles (Lord and Beevers, 1972). It remains to be seen whether the peroxisomal membrane is impermeable to either NADH/NAD or hydroxypyruvate/glycerate, or to both the coenzyme and metabolite.

The glycolate pathway in photorespiration generates NAD and the reverse glycerate pathway produces NADH. NADH oxidase has not been found in the peroxisomes. To account for the reoxidation of NADH or the reduction of NAD, an electron shuttle system has been proposed (Fig. 4.6). The existence of a highly active peroxisomal NAD-malate dehydrogenase, which has no other known metabolic function in the organelles, and a peroxisomal aspartate-α-ketoglutarate aminotransferase were invoked in the shuttle mechanism. Malate and aspartate, but not oxaloacetate, would be the vehicles of the shuttle. This proposed shuttle system is similar to the proven system in the mitochondria and the proposed system in glyoxysomes (Section IV,D).

The leaf peroxisomal malate dehydrogenase and aspartate-α-ketoglutarate aminotransferase exist as isozymes (Table 4.4) that are distinct from those of the mitochondria, cytosol, or plastids (see review by Ting *et al.*, 1975). The respective isozymes of either enzyme are different in their electrophoretic mobility, immunological kinship, and kinetic properties. It is interesting that the spinach leaf peroxisomal malate dehydrogenase and the castor bean glyoxysomal malate dehydrogenase have very similar properties (Huang *et al.*, 1974). The two enzymes occur in two different types of peroxisomes of different tissues and species, and yet they share properties that are more similar between themselves than among the other malate dehydrogenase isozymes in each respective tissue, such as immunological kinship, high sensitivity toward heat inactivation, and low electrophoretic mobility at neutral pH. This similarity suggests a close genetic relationship between the two enzymes and thus the two types of peroxisomes.

D. Peroxisomes in Leaves of C_4 Species

C_4 species are plants that eliminate the apparent photorespiration and thus are more efficient in photosynthesis. Although all C_4 species carry out a similar initial fixation of CO_2 into a C_4 acid, they differ widely in their mechanisms of eliminating the apparent photorespiration. Depending on the species, the leaves may contain high or low activities of the glycolate pathway enzymes (Table 4.7). Peroxisomes are observed in the mesophyll and bundle sheath cells of all C_4 species, and they contain catalase as indicated by the diaminobenzidine staining method (Chapter 2). Since leaf peroxisomes are major subcellular sites of photorespiration in C_3 species, their functional roles in the leaves of C_4 plants are of interest.

The details of C_4 photosynthesis will not be discussed but can be obtained from many excellent reviews (e.g., Edwards and Huber, 1981). In the current discussion on peroxisomes in leaves of C_4 species, the C_4 species will be divided into two groups according to their content of high or low activities of the glycolate pathway enzymes. It should be

TABLE 4.7

Amounts or Protein, Chlorophyll, and Enzyme Activities Recovered per Gram Fresh Weight of the Various Plant Tissues[a]

	Spinacia oleracea (C_3)	Atriplex patula (C_3)	Atriplex rosea (C_4)	Sorghum sudanense (C_4)
Protein (mg)	41	21	37	28
Chlorophyll (mg)	1.2	—[b]	2.0	2.1
RuBP carboxylase	12.2	—	3.1	5.2
PEP carboxylase	1.0	—	18.5	14.2
Malic enzyme	—	—	—	24.7
Cytochrome oxidase	1.5	—	2.4	1.4
Catalase	1638	1980	958	539
Hydroxypyruvate reductase	52.0	21.3	9.2	1.6
Glycolate oxidase	2.12	0.95	0.53	0.08

[a] *Atriplex rosea* and *Sorghum sudanense* represent the extremes of C_4 species with high and low, respectively, activities of the glycolate pathway enzymes. Enzyme activities are expressed as μmol/min/g fresh weight. Data are from Huang and Beevers (1972).

[b] Amount not tested.

understood that there are many C_4 species possessing intermediate characteristics, and that the expression of the characteristics within a species may be influenced by environmental factors.

Those C_4 species with high activities of glycolate pathway enzymes included in the so-called "phosphoenolpyruvate carboxykinase type" and "NAD-malic enzyme type" (Edwards and Huber, 1981). The glycolate pathway enzymes are localized together with RuBP carboxylase and other Calvin cycle enzymes in the bundle sheath cells. There, glycolate is produced and metabolized in a manner similar to that in the mesophyll cells of C_3 species. However, before the photorespired CO_2 produced in the bundle sheath cells escapes into the atmosphere, it is refixed in the surrounding mesophyll cells by the active phosphoenolpyruvate carboxylase. Thus, no overt photorespiration is observed.

The C_4 species with low activities of glycolate pathway enzymes in the leaves are the "NADP-malic enzyme type" (Edwards and Huber, 1981). In the bundle sheath cells, the chloroplasts have greatly reduced photosystem II activity such that the photosynthetic O_2 production is severely limited, and little O_2 is available for glycolate formation by RuBP carboxylase. Therefore, insignificant amounts of glycolate are produced, and there are low activities of the glycolate pathway enzymes.

Thus, it seems that peroxisomes with leaf-type characteristics should be abundant in, and perhaps restricted to, the bundle sheath of the C_4 species having high glycolate pathway enzyme activities. However, direct experimental evidence is lacking. Although counting the number of peroxisomes in the bundle sheath cells and in the mesophyll cells has been carried out on many C_4 species (Chapter 2), no account has been made on the number of peroxisomes in the entire mesophyll tissue compared to the entire bundle sheath tissue. Furthermore, a detailed comparative study between the two groups of C_4 species has not been reported.

Peroxisomes have been isolated on sucrose gradients from the leaves of several C_4 species (Chapter 2). Limited information is available on their characteristics other than their content of catalase and glycolate oxidase. However, the existence of these two enzymes is not a characteristic of leaf peroxisomes; rather, they occur in all types of peroxisomes. Whether the peroxisomes in the leaves of C_4 species are truly "leaf-type" peroxisomes is unknown. Just as we clearly know that green chloroplasts in C_4 species do not necessarily contain the complete Calvin cycle, the peroxisomes in the leaves of C_4 species may not possess active glycolate pathway enzymes typifying the highly specialized C_3 leaf peroxisomes. They may simply be unspecialized peroxisomes of the type found in many nongreen and nonoil tissues. It seems senseless to have leaf-type peroxisomes containing the glycolate pathway enzymes in the mesophyll cells where glycolate is not produced. In the leaves of C_4 species, whenever active glycolate pathway enzymes are present, they are localized in the bundle sheath rather than in the mesophyll, suggesting that the leaf-type peroxisomes, if existing, are present or at least concentrated in the bundle sheath (Huang and Beevers, 1972; Gerhardt, 1978). As judged from a compilation of information on peroxisomes isolated from various higher plant tissues, leaf peroxisomes have a ratio of glycolate oxidase activity to catalase activity much higher than those of other types of peroxisomes. Based on this observed ratio in the mesophyll extracts and bundle sheath extracts of C_4 species, it has been suggested that, in *Sorghum sudanense* and perhaps other C_4 species too, the mesophyll peroxisomes are not the highly specialized leaf-type peroxisomes but are only unspecialized peroxisomes like those found in many nongreen tissues (Huang and Beever, 1972; Gerhardt, 1978; Edwards and Huber, 1981). The term "leaf" peroxisome with its functional meaning cannot be applied to all the peroxisomes in the leaves of C_4 species.

The CAM (crassulacean acid metabolism) plants are related to the C_4 plants in that they also initially fix CO_2 into a C_4 acid under special environmental conditions (Kluge and Ting, 1978). However, they only have one type of leaf cells possessing chloroplasts, and carry out CAM metabolism or normal C_3 photosynthesis and photorespiration depending on the environmental conditions. Peroxisomes in the leaves of CAM species have been observed with the electron microscope, and the organelles also have been isolated on sucrose gradients (Chapter 2). Although the isolated peroxisomes possess catalase and glycolate oxidase, the other properties of the organelles have not been investigated. It is assumed that they are the typical leaf-type peroxisomes of C_3 species.

Most of our current information on leaf peroxisomes is based mainly on the studies of the organelles from spinach leaves. It is surprising that intensive study has not been extended to other species. Especially so is the lack of information on the peroxisomes in the leaves of C_4 and CAM species. A careful comparative study of leaf peroxisomes from various species would be warranted.

VI. PEROXISOMES IN UREIDE METABOLISM

The ureides, allantoin and allantoic acid, are important metabolites for nitrogen transport in the xylem of some plant species. The ureides have the advantage over other nitrogen-transporting metabolites such as asparagine or glutamine in having the highest nitrogen to carbon ratio. In species of the genera *Acer, Platanus,* and *Aesculus,* the xylem sap contains allantoin and allantoic acid as major forms of nitrogen, representing 10–99% of the total nitrogen (Thomas and Schrader, 1981). For the transport of organic nitrogen from the nodules to the shoots in symbiotic nitrogen fixing legumes, some species like *Vicia* and *Pisum* use asparagine whereas other species like *Glycine* and *Vigna* utilize ureides as the major nitrogen carriers (review by Rawsthorne *et al.,* 1980). In plant species utilizing ureides as the nitrogen carriers, the metabolism of ureides in the supply and receiving tissues is of great physiological importance.

The metabolic pathway of ureides is shown in Fig. 4.8. In some ureotelic animals, several of the enzymes, including xanthine oxidase (dehydrogenase), urate oxidase, and allantoinase, are localized in peroxisomes (Tolbert, 1981). In plants, urate oxidase is present in all types of peroxisomes (Chapter 2). In some of these peroxisomes, allantoinase also is present. However, unlike those in ureotelic animals, the specific activities of urate oxidase and allantoinase in plant peroxisomes generally are very low (Tables 4.1 and 4.2), and the enzymes have been assumed simply to be characteristic peroxisomal enzymes of little physiological importance. The peroxisomes in the tissues of those plant species that utilize ureides as nitrogen carriers have only been studied recently.

In nodule extracts of *Glycine* (soybean) and *Vigna* (cowpea), substantial activities of xanthine dehydrogenase, urate oxidase, and allantoinase are present (Rawsthorne *et al.,* 1980). In soybean nodules, large and numerous peroxisomes are present in the uninfected root cells adjacent to the infected root cells which contain only small or degenerated peroxisomes (Newcomb and Tandon, 1981) (Fig. 4.9). In the nodules, catalase and urate oxidase are localized in the peroxisomes, whereas xanthine oxidase and allantoinase are restricted to the cytosol and the microsomes, respectively (Hanks *et al.,* 1981). Presumably, allantoin is produced by urate oxidase in the peroxisomes in the uninfected nodule cells from where it moves to the xylem cells. Peroxisomes isolated from soybean nodules contain a high activity of urate oxidase which is at least severalfold higher than that in any other type of plant peroxisomes (Table 4.1). In soybean leaves, little allantoicase is present; catalase, together with only a trace amount of urate oxidase, exists in the leaf peroxisomes, whereas allantoinase occurs in the microsomes. Apparently, unlike the peroxisomes in some ureotelic animals, soybean nodule peroxisomes contain only one enzyme of the ureide metabolic pathway. Nevertheless, the metabolism of ureide in relation to nitrogen

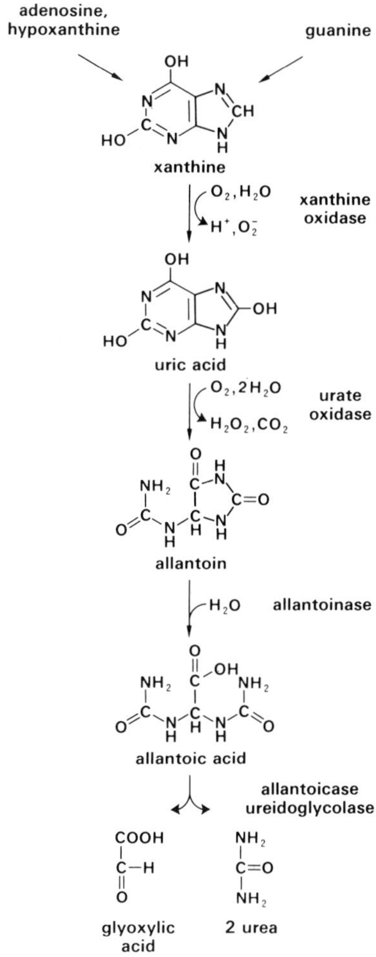

Fig. 4.8. Ureide metabolic pathway. In the last step, allantoic acid is metabolized first to urea and ureidoglycolic acid by allantoicase, and subsequently ureidoglycolic acid is converted to glyoxylic acid and urea by ureidoglycolase. In some microorganisms, the last step involves the conversion of allantoic acid to ureidoglycolic acid, CO_2, and two ammonia by allantoate amidohydrolase, and ureidoglycolic acid is converted to glyoxylic acid and urea by ureidoglycolase. See review by Thomas and Schrader (1981).

transport should be recognized as an important physiological role carried out by plant peroxisomes (Table 4.8).

Indirect evidence for the involvement of peroxisomes in ureide metabolism related to ureide translocation is present. When roots of *Phaseolus coccineous* are grown in elevated nitrogen salts, activities of urate oxidase

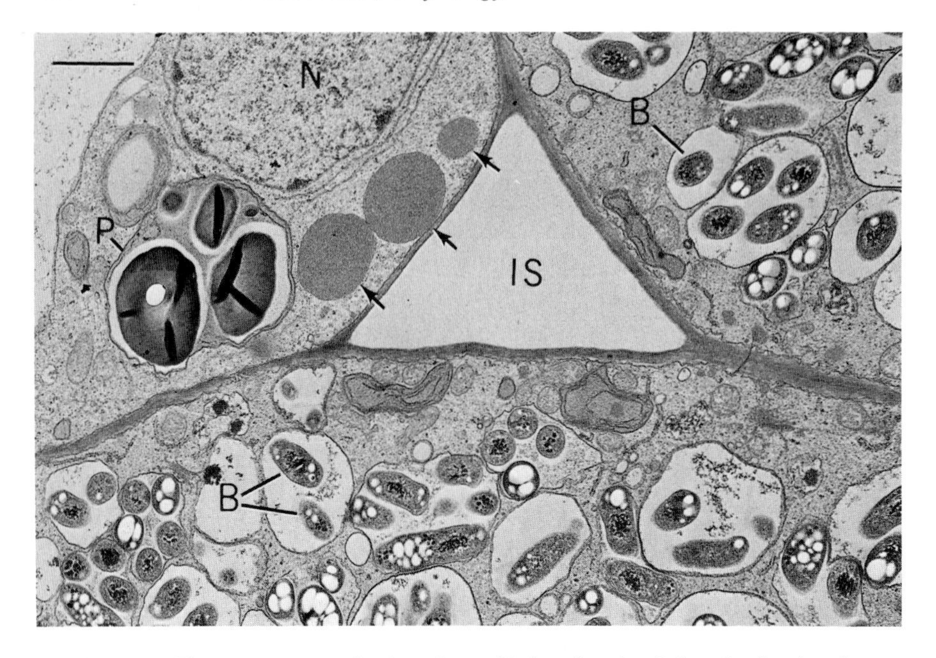

Fig. 4.9. Electron micrograph of portions of infected and uninfected cells of soybean nodule, showing (by arrows) several microbodies (peroxisomes involved in ureide metabolism) in the uninfected cell. At bottom and upper right are infected cells with numerous bacteroids (B) in vacuoles. IS, intercellular space; N, nucleus; P, plastid with several starch grains. Bar = 1 μm. From Newcomb and Tandon (1981).

and catalase in crude extracts and in isolated peroxisomes increase (Theimer and Heidinger, 1974). Presumably, the nitrogen salts are assimilated into purines, which are oxidized in the peroxisomes to produce allontoin, the transport form of nitrogen in the bean plants. In mustard cotyledons at the late stage of postgermination, peroxisomes with higher activity of urate oxidase appear (Hong and Schopfer, 1981; also see Fig. 6.8). It is possible that these peroxisomes are involved in the production of allantoin from purines as a part of the translocation process in a senescencing organ.

VII. PEROXISOMES IN FUNGI AND ALGAE

Peroxisomes have been observed by electron microscopy in numerous fungi and algae, regardless of whether they are autotrophs or heterotrophs growing on diverse media (Chapter 2). The literature is vast, and contains considerable confusion. A review has not appeared which puts our current knowledge of the physiological aspect of these peroxisomes in proper perspectives.

There are technical difficulties inherent in studies of peroxisomes in microorganisms. Algal or fungal cells are usually difficult to break gently, and therefore fragile organelles such as peroxisomes are disrupted easily during cell fractionation. This difficulty may be a major cause for the varying reports of enzyme localization, such as in *Saccharomyces cerevisiae*. Wall-less mutants or preparation of protoplasts before cell lysis have been employed with only limited success. Still, the general trend is that good organelle preparations have been generated from a few species, whereas isolations from other species always have met with failure. Besides the problem with cell disruption, other inherent difficulties exist. Catalase and isocitrate lyase have been two enzymes commonly employed as markers for peroxisomes and the glyoxylate cycle, respectively. In *Saccharomyces cerevisiae*, two forms of catalase exist, one localized in the peroxisomes and the other in the vacuole (Susani *et al.*, 1976). Therefore, catalase cannot be used as the sole marker for peroxisomes in this yeast. Isocitrate lyase is an inducible enzyme when microorganisms are grown on acetate, ethanol, or alkane, and those microorganisms grown on glucose or sucrose contain a low activity of the enzyme. In *Neurospora*, two isozymic forms of isocitrate lyase are present; one is produced constitutively in a low amount and the other is the inducible enzyme (Sjogren and Romano, 1967). Using isocitrate lyase as a marker for glyoxylate cycle or glyoxysome-like organelles is not neces-

TABLE 4.8

Various Types of Peroxisomes in Higher Plants, Fungi, and Algae Catagorized according to Metabolic Roles[a]

Higher plants	Fungi and algae
1. "Unspecialized" peroxisomes	1. "Unspecialized" peroxisomes
2. Glyoxysomes	2. Glyoxysome-like peroxisomes
3. Leaf peroxisomes	3. Peroxisomes in photorespiratory glycolate metabolism
4. Peroxisomes in ureide metabolism	4. Peroxisomes in ureide metabolism
	5. Peroxisomes in methanol oxidation
	6. Peroxisomes in oxalate synthesis
	7. Peroxisomes in amine metabolism

[a] The term "unspecialized" is used only tentatively to describe those peroxisomes with no recognized major physiological roles. Whereas peroxisomes of similar metabolic roles are found in both higher plants and fungi/algae (the first four), peroxisomes for methanol oxidation, oxalate synthesis, and amine metabolism are reported only in fungi/algae.

sarily valid, and peroxisomes containing isocitrate lyase in glucose-grown microorganisms cannot be viewed as organelles equivalent metabolically to glyoxysomes.

With few exceptions, the microbial peroxisomes migrate to an equilibrium density greater than that of the mitochondria during sucrose gradient centrifugation. In those species where the peroxisomes have been isolated and studied, the metabolic roles of the organelles have been assessed. At least seven types of peroxisomes can be distinguished according to their metabolic roles (Table 4.8). It is anticipated that peroxisomes with other metabolic roles will be identified in the future, such as peroxisomes in those microorganisms grown on glycolate or D-amino acid as the sole carbon and/or nitrogen source.

A. Peroxisomes of Unknown Function: "Unspecialized" Peroxisomes

Peroxisomes are present in most if not all fungi and algae when examined with the electron microscope (Chapter 2). In many of these studies, the content of catalase in the peroxisomes has been confirmed by the diaminobenzidine cytochemical method, and in a few studies by direct enzyme assay on organelles isolated on sucrose gradients (Chapter 2). Whereas the peroxisomes in microorganisms grown under certain conditions perform unique physiological functions (Table 4.8), the peroxisomes in other microorganisms, especially those grown on glucose, sucrose, or lactose as the sole carbon source, are not known to carry out specific active metabolic roles. In these microorganisms, the number of peroxisomes per cell, or the number compared with that of other organelles, is generally small. No physiological function has been proposed for these peroxisomes other than a general detoxication of H_2O_2. In essence, they are similar to the "unspecialized" peroxisomes of higher plants. Again, caution should be used since the term "unspecialized" refers to the lack of a known special physiological role, and the status may be altered with new information.

B. Glyoxysome-Like Peroxisomes in Heterotrophic Fungi and Algae

When heterotrophic fungi or algae are grown on carbon sources that are the same or related to the metabolites generated during the early steps of the gluconeogenic pathway from triacylglycerols in oilseeds, synthesis of glyoxysome-like peroxisomes and their related enzymes are induced. These carbon sources, including alkane, fatty alcohols, fatty acids, ethanol, and acetate, can enter the known metabolic pathway directly or through initial modification (Fig. 4.10). Also, the spores of some fungi contain storage lipid, which is mobilized during or following spore germination; this situation appears similar to the germination of oilseeds.

Most of the investigations on glyoxylate cycle metabolism in heterotrophic fungi and algae are not sufficient for a complete understanding of the whole process, e.g., initial modification of substrates and their subcellular localization. Interpretation must rely heavily on our knowledge of glyoxylate cycle metabolism in oilseeds and known enzymatic reactions in bacterial systems. Nevertheless, information is available to indicate that glyoxylate cycle metabolism in heterotrophic fungi and algae is similar to that in oilseeds, but with differences. The induction of the enzymes by the various substrates in fungi has been reviewed (Maxwell *et al.*, 1977). In this section, the peroxisomal metabolism will be emphasized.

A major difference between the glyoxysome-like peroxisomes in fungi and the glyoxysomes in oilseeds is that the fungal peroxisomes contain only the two key enzymes of the glyoxylate cycle, malate synthase and isocitrate lyase,. and thus operation of the complete glyoxylate cycle requires sharing enzymes with the Krebs cycle in the mitochondria (Table 4.9). Among the heterotrophic algae, only *Euglena gracilis* grown on ethanol or hexanoic acid is known to have enzymes of the complete

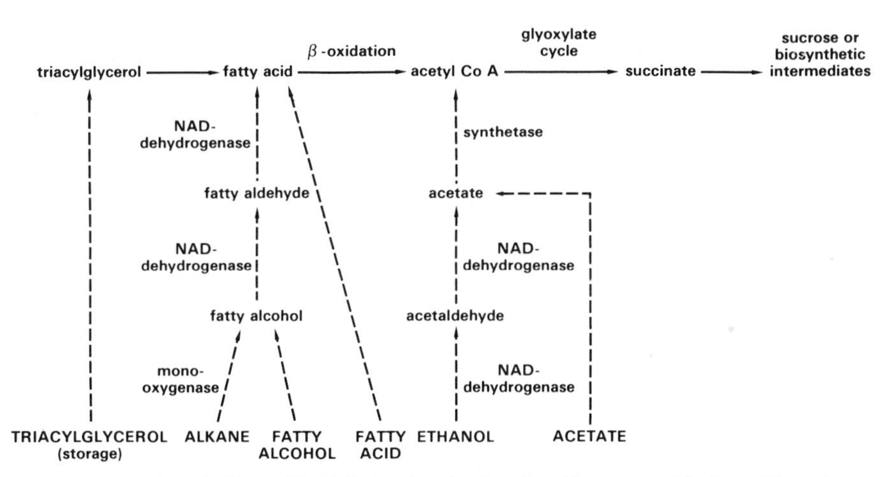

Fig. 4.10. Metabolism of lipid derivatives by fungi and heterotrophic algae. The microorganisms convert the lipid derivatives (shown in capital letters on the lower side) to initial products that can be metabolized by a metabolic pathway (shown on the upper side) known to exist in oilseeds.

TABLE 4.9

Enzyme Contents in Glyoxysome-Like Peroxisomes Isolated by Gradient Centrifugation from Heterotrophic Fungi and Algae Using Lipid, Ethanol, or Acetate as the Sole Carbon Source[a]

Carbon source	Species	Catalase	Urate oxidase	D-amino acid oxidase	Isocitrate lyase	Malate synthase	Malate dehydrogenase, citrate synthase, aconitase	β-Oxidation	References
Internal lipid	*Blastocladiella emersonii*	+			+	+			Mills and Cantino (1975)
	Botryodiplodia theobromae	+			+	+	−	−	Armentrout and Maxwell (1981)
	Entophylyctis variabilis	+			+	+			Powell (1976)
Alkanes	*Candida tropicalis*	+	+	+	+	+	−	+	Kawamoto *et al.* (1977)
Hexanoate	*Euglena gracilis*	−			+	+	+	+	Graves and Becker (1974)
Ethanol/acetate	*Aspergillus tamarii*	+			+	+	−		Graves *et al.* (1976)
	Coprinus lagopus				+	+			Sullivan and Casselton (1972)
	Saccharomyces cerevisiae	+			+	+	−		Szab and Avers (1969)
	Neurospora sp.	+	+	+	+	+			Kobr *et al.* (1973) Theimer *et al.* (1978)
	Euglena gracilis	−			+	+	+		Graves *et al.* (1972)
	Polytomella caeca	+	+		−	−			Gerhardt (1971), Cooper and Lloyd (1972)

[a] + and − denote the presence and absence, respectively, of activities examined. *Candida* peroxisomes also contain fatty alcohol dehydrogenase, fatty aldehyde dehydrogenase, fatty acyl-CoA thiokinase, creatine acetyltransferase, NAD-glycerolphosphate dehydrogenase, and NADP-isocitrate dehydrogenase that have not been reported in the peroxisomes of the other species listed. In *Neurospora* grown on acetate, catalase and urate oxidase might be present in one type of peroxisomes and isocitrate lyase and malate synthase in another type of peroxisomes (Theimer *et al.*, 1978).

glyoxylate cycle in the peroxisomes (Graves *et al.*, 1972; Graves and Becker, 1974). Since no similar studies have been made on other algal species, it is not known whether *Euglena* is unique among the heterotrophic algae.

1. Peroxisomes in Fungal Spores with Storage Triacylglycerol

The spores of some fungi contain storage lipid in lipid bodies. Following germination, many peroxisomes are situated adjacent to the lipid bodies (micrographs in Chapter 2). The peroxisomes have been isolated from three fungal species. In *Blastocladiella emersonii*, the zoospores contain smaller peroxisomes, which eventually fuse to form larger peroxisomes (symphyomicrobody) in the mycelium (Mills and Cantino, 1975). On sucrose gradients, the small peroxisomes sediment at an equilibrium density of 1.22 g/cm^3 and the large peroxisomes at 1.29 g/cm^3. The peroxisomes contain catalase, isocitrate lyase, and malate synthase (Table 4.9). Peroxisomes of similar enzyme content also have been isolated from the zoospores of *Entophylyctic variabilis* (Powell, 1976). More information is available on the enzyme content of peroxisomes isolated from the germinating conidia of *Botyodiplodia theobromal* (Armentrout and Maxwell, 1981). The peroxisomes, sedimented at an equilibrium density of 1.22 g/cm^3, contain catalase, isocitrate lyase, and malate synthase. Citrate synthase and malate dehydrogenase are present in the mitochondria and not in the peroxisomes. Of the β-oxidation enzymes tested, enoyl-CoA hydratase is present only in the mitochondria, and about 80% of thiolase and β-hydroxyacyl-CoA dehydrogenase activities are present in the mitochondria with the remainder in the peroxisomes. It appears that the peroxisomes in the spores of the above three fungal species are not as enzymatically complex as those in the oilseeds, since they have to rely heavily on the cooperation with the mitochondria to carry out the conversion of fatty acid to acetate. At least in *Botryodiplodia theobromal*, the mitochondria can carry out active β-oxidation of fatty acid to acetate independent of the peroxisomes.

2. Peroxisomes in Fungi Grown on Alkane, Fatty Alcohol, and Fatty Acid

Some fungi and bacteria can use alkane, fatty alcohol, or fatty acid as the sole carbon source for growth. There is an increasing interest in this research area because of its applicability to converting petroleum byproducts into consumable foodstuffs, cleaning accidental oil spills, and resolving corrosion problem in jet aircraft fuel systems due to fungal growth.

The initial enzymatic steps for the utilization of alkane in fungi (*Can-*

dida tropicalis and *Cladosporium resinae*) and bacteria involve the following (LeBault *et al.*, 1970; Walker and Cooney, 1973; Lode and Coon, 1973):

$$\underset{\text{Alkane}}{R\text{—}CH_3} \xrightarrow[\text{mono-oxygenase}]{\text{NAD, } O_2} \underset{\text{Alcohol}}{R\text{—}CH_2OH} \xrightarrow{\text{NAD}} \underset{\text{Aldehyde}}{R\text{—}CHO} \xrightarrow{\text{NAD}} \underset{\text{Acid}}{R\text{—}COOH} \rightarrow \beta\text{-oxidation} \quad (4.10)$$

The first enzyme is monooxygenase requiring both O_2 and NAD as electron acceptors. The second and third enzymes presumably utilize NAD as an electron acceptor. Recently, the possibility has been raised that the microbial alcohol oxidizing enzyme (from alcohol to aldehyde) actually may utilize O_2 instead of NAD as the electron acceptor; the enzyme in jojoba cotyledons storing wax ester employs O_2 and not NAD (Moreau and Huang, 1979). The resultant fatty acid is oxidized by the β-oxidation system to acetate.

The involvement of peroxisomes in assimilation of hydrocarbon derivatives has been well studied in *Candida tropicalis* (Mishina *et al.*, 1978; Kawamoto *et al.*, 1979; Yamada *et al.*, 1980; Tanaha *et al.*, 1982). The peroxisomes and related enzymes are induced when the yeast is grown on alkane, and the peroxisomes become more numerous than the mitochondria (Fig. 4.11). The peroxisomes can be well separated from the mitochondria on a sucrose gradient (Fig. 4.12). The monooxygenase for the oxidation of alkane to alcohol is located in the microsomes. About one-third of the cellular fatty alcohol dehydrogenase, fatty aldehyde dehydrogenase, and fatty acyl-CoA synthetase, is present in the mitochondria and microsomes, presumably for lipid biosynthesis. The remaining two-thirds of all three enzyme activities are present in the peroxisomes, which also contain the characteristic peroxisomal enzymes catalase, urate oxidase, and D-amino acid oxidase. In addition, the peroxisomes contain all the cellular enzymes of the fatty acid β-oxidation system as well as isocitrate lyase and malate synthase (Table 4.9). Citrate synthase, malate dehydrogenase, and aconitase are absent from the peroxisomes. The mitochondria do not contain the β-oxidation enzymes, but possess enzymes of the complete Krebs cycle. Thus, the peroxisomes and the mitochondria must cooperate in operation of the glyoxylate cycle, with metabolites shuttling between the two organelles. In addition, both organelles contain carnitine acetyltransferase; this enzyme can provide a mechanism for transferring acetate units generated by the β-oxidation system in the peroxisomes to the mitochondria for the citrate synthase reaction (Fig. 4.13). Most of the cellular NAD-glycerol phosphate dehydrogenase and FAD-glycerol phosphate dehydrogenase are restricted to the peroxisomes and the mitochondria,

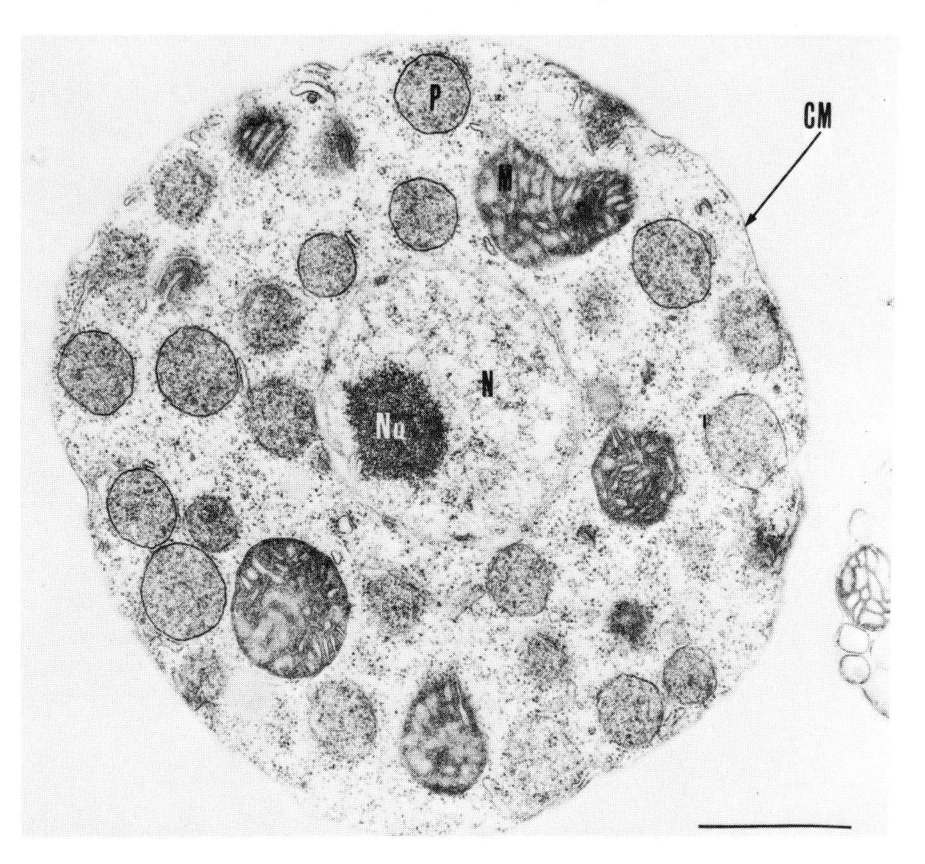

Fig. 4.11. Electron micrograph of a protoplast of *Candida tropicalis* grown on *n*-alkane mixture (C_{10}–C_{13}) as the sole carbon source. Bar = 1 μm. CM, cytoplasmic membrane; M, mitochondrion; P, peroxisome, N, nucleus; Nu, nucleolus. From Tanaka *et al.* (1982).

respectively. The two enzymes can provide an electron shuttle system for transferring the NADH generated by the β-oxidation in the peroxisomes to the mitochondria for oxidation, using glycerol phosphate and dihydroxyacetone phosphate as the vehicles. Such an electron shuttle system is known to occur between the cytosol and the mitochondria in mammalian cells.

The assimilation pathway of alkane in *Candida* is quite similar to the gluconeogenic pathway in jojoba seeds, which store wax ester instead of triacylglycerol as a food reserve (Moreau and Huang, 1977). Although having many similarities, the *Candida* peroxisomes differ from the oilseed glyoxysomes in several aspects. The yeast peroxisomes do not contain a complete glyoxylate cycle, but possess carnitine acetyl-

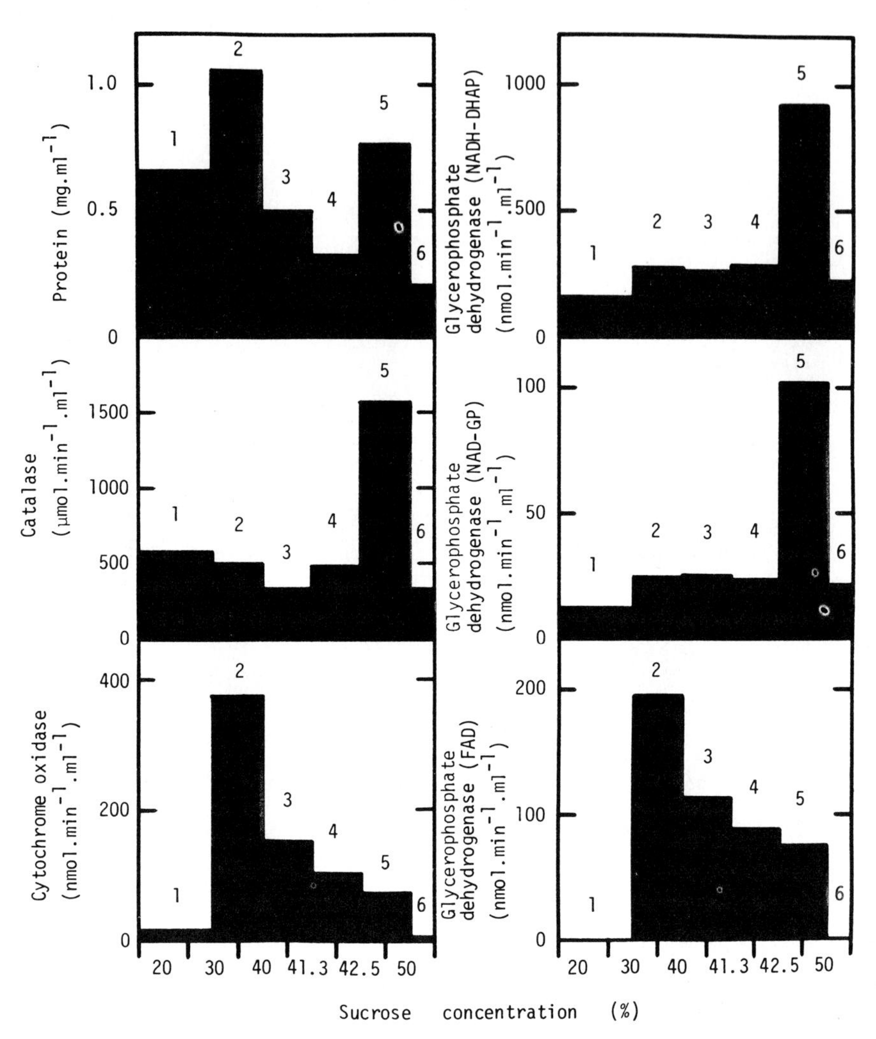

Fig. 4.12. Separation of organelles from a particulate fraction (pellet obtained after centrifugation at 20,000 *g*) of extract from *Candida tropicalis* grown on alkanes. Catalase and cytochrome oxidase are markers of peroxisomes and mitochondria, respectively. From Kawamoto *et al.* (1979).

transferase and NAD-glycerol phosphate dehydrogenase. These two enzymes are unknown in the peroxisomes of higher plants but exist in mammalian peroxisomes (Tolbert, 1981). The presence of fatty alcohol dehydrogenase and fatty aldehyde dehydrogenase is unique to the yeast peroxisomes. Although the two enzymes (fatty alcohol oxidase and fatty aldehyde dehydrogenase) also are present in jojoba seeds, they are localized mainly in the wax bodies with little in the glyoxysomes (Moreau and Huang, 1979).

In heterotrophic *Euglena* grown on hexanoate, the isolated peroxisomes contain fatty acyl CoA synthetase, both of the β-oxidation enzymes examined, and all the glyoxylate cycle enzymes (Graves and Becker, 1974). The *Euglena* peroxisomes are thus very similar to the glyoxysomes of higher plants. No catalase is present in these peroxisomes or in the algal extract. In the absence of catalase, it remains to be seen if the peroxisomal fatty acyl CoA oxidase is still a flavin protein generating H_2O_2 or a dehydrogenase such as the mitochondrial enzyme in mammals.

3. Peroxisomes in Fungi and Algae Grown on Ethanol or Acetate

When fungi and heterotrophic algae use ethanol or acetate as the sole carbon source for growth (Fig. 4.10), an increase in enzyme activities that can convert ethanol or acetate into carbohydrate is observed. Ethanol is first converted to acetate by ethanol dehydrogenase and acetaldehyde dehydrogenase. Acetate is activated to acetyl-CoA, which is then channeled through the glyoxylate cycle. The two dehydrogenases and the acetyl CoA synthetase (Fig. 4.10) are induced by ethanol

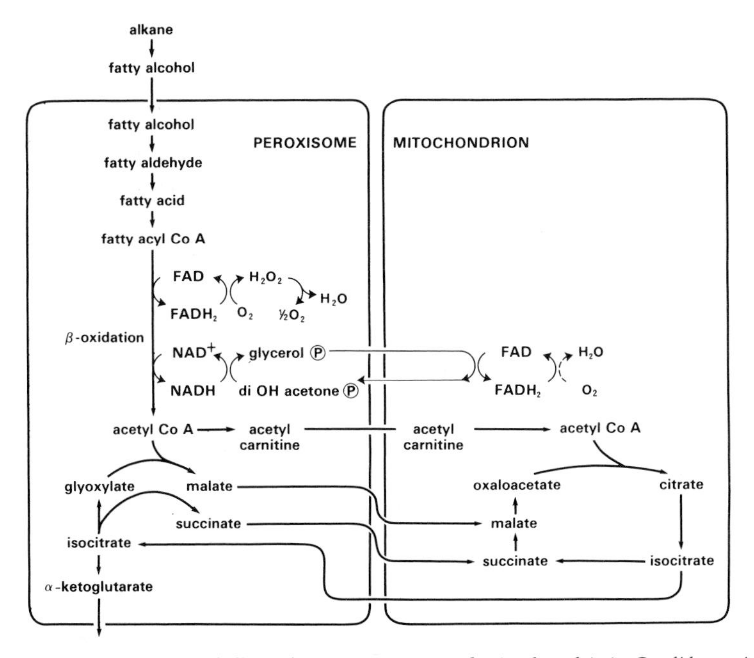

Fig. 4.13. Metabolism of alkane by peroxisomes and mitochondria in *Candida tropicalis*.

and acetate in *Euglena* (Graves and Becker, 1974; Woodward and Merrett, 1975).

The peroxisomes isolated on sucrose gradients from several fungal species contain the characteristic peroxisomal enzymes, including catalase, urate oxidase, and D-amino acid oxidase (Table 4.9). In addition, the peroxisomes possess malate synthase and isocitrate lyase but not the other three enzymes (citrate synthase, aconitase, and malate dehydrogenase) of the glyoxylate cycle. Thus, the peroxisomes and the mitochondria must work together for operation of the glyoxylate cycle. The mechanism of cooperation, such as the metabolite shuttle and source of acetyl-CoA in each organelle, is unknown. Whether the mechanism is similar to that of *Candida* grown on alkane (Fig. 4.13) remains to be seen.

Two types of peroxisome-like organelles may exist in slime mutants of *Neurospora crassa* grown on acetate as the sole carbon source (Theimer *et al.*, 1978; Wanner and Theimer, 1982). When the fungal organelles were separated on a sucrose gradient, mitochondria appeared at a density of 1.18 g/cm^3, whereas malate synthase and isocitrate lyase occurred at a density of 1.21 g/cm^3 and catalase and urate oxidase at a density of 1.24 g/cm^3 (Fig. 4.14). Electron microscopy of gradient fractions or intact cells also suggested the existence of two types of microbodies, only one of which showed positive staining with diaminobenzidine for catalase activity. However, the diaminobenzidine-positive and diaminobenzidine-negative organelles have not been related directly to the gradient fractions containing catalase (and urate oxidase) and malate synthase (and isocitrate lyase), respectively. The catalase- and urate oxidase-containing organelles were present in *Neurospora* grown on either acetate or sucrose, and appear to be the "unspecialized" peroxisomes described in Section VII,A. An earlier study on *Aspergillus* showed very similar findings, but the results were interpreted differently (Graves *et al.*, 1976). When the organelles in extract of ethanol-grown *Aspergillus* were separated by sucrose density gradient centrifugation, malate synthase and isocitrate lyase peaked at a density of 1.23 g/cm^3 together with a small shoulder of catalase, which peaked at a density of 1.26 g/cm^3 (a similar catalase shoulder, though very minor, also is present in the *Neurospora* glyoxysome-like peroxisomes in Fig. 4.14). These workers suggested that the glyoxysome-like peroxisomes contained malate synthase, isocitrate lyase, and catalase, and that the catalase peak at a higher density belonged to another particle (microbody inclusions or the Woronin bodies).

It remains to be seen if the existence of two types of peroxisomes in *Neurospora* (perhaps also in other fungi) is valid. If so, it is interesting

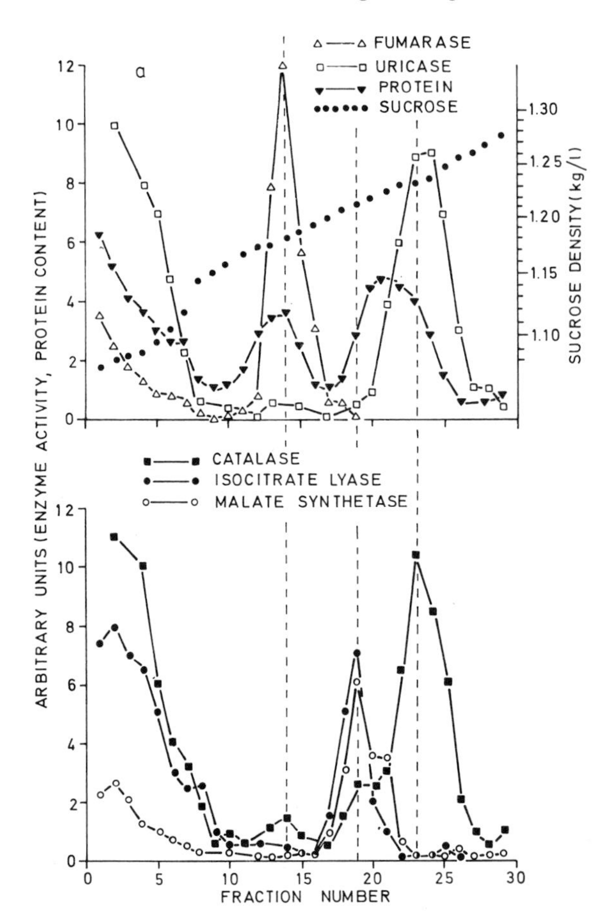

Fig. 4.14. Separation of organelles from crude extract of *Neurospora crassa* grown in sucrose medium for 30 hr and then in acetate medium for 6 hr. Note the separation of mitochondria (fumarase as marker), organelles containing isocitrate lyase and malate synthase, and peroxisomes possessing catalase and urate oxidase. From Theimer *et al.* (1978).

that the *Neurospora* (or fungal) glyoxysome-like peroxisomes do not contain catalase. In *Euglena* grown on fatty acid (Section VII,B,2) or acetate (following paragraph), the glyoxysome-like peroxisomes also are devoid of catalase (as are the peroxisomes in autotrophic *Euglena;* see Section VII,C). Strictly speaking, the term "peroxisome" cannot be used to describe the glyoxysome-like peroxisomes in *Neurospora* or *Euglena.* Yet, these organelles are closely related morphologically and enzymically to the glyoxysomes in oilseeds and glyoxysome-like peroxisomes in other fungi. Caution should be taken in accepting this apparent irregularity, since the presence of catalase in the *Neurospora* (preceding paragraph) and *Euglena* peroxisomes has been suggested or reported. In *Euglena*, despite several laboratories reporting the absence of catalase in the peroxisomes, it was shown clearly (Brody and White, 1972; White and Brody, 1974) that the peroxisomes in *Euglena* grown on glucose or acetate, as well as those grown aerobically (but not those grown in CO_2-free air), possessed catalase. These positive results were obtained using both the 3,3'-diaminobenzidine cytochemical technique on intact cells and isolated peroxisomes, and a direct spectrophotometric enzyme activity assay on cell-free extracts and isolated peroxisomes.

In heterotrophic algae grown on acetate or ethanol, the subcellular localization of the glyoxylate cycle may be species-dependent. In *Euglena gracilis* (Graves *et al.*, 1972), the isolated peroxisomes contain the complete glyoxylate cycle enzymes (all five enzymes except aconitase were studied). No catalase is present in the peroxisomes or in the algal extract, as is the case in autotrophic *Euglena* (see Section VII,C). In a separate investigation, malate synthase and isocitrate lyase were also found to be particulate in *Euglena gracilis* and *Astasia longa*, but an attempt to separate the peroxisomes from the mitochondria in the particulate fraction was unsuccessful (Begin-Heick, 1973). On the contrary, peroxisomes isolated from *Polytomella caeca* (a wall-less, single cell, green algae) contain catalase and urate oxidase but no malate synthase and isocitrate lyase; the latter enzymes exist in the soluble fraction (Gerhardt, 1971; Cooper and Lloyd, 1972). In view of the possible existence of two types of peroxisomes in *Neurospora* (preceding paragraph), it is still possible that in *Polytomella* cells there is an additional type of peroxisome that contains the complete glyoxylate cycle enzymes but no catalase (like the *Euglena* peroxisomes) and that is relatively more fragile and is broken during organelle preparation. The uniqueness of the *Euglena* peroxisomes, or the *Polytomella* peroxisomes, among diverse algal species cannot be ascertained because there are so few reports on successful peroxisome isolation from algae.

C. Peroxisomes and Photorespiration in Autotrophic Algae

1. Photorespiration and Glycolate Metabolism in Algae

In many aspects, algae carry out photorespiration in a way similar to that in higher plants. The similarities include their being subject to the Warburg effect of oxygen inhibition of photosynthesis, the properties of ribulose bisphosphate carboxylase/oxygenase, their possession of phosphoglycolate phosphatase and the glycolate pathway enzymes, the general patterns of metabolism of radioactive CO_2 or glycolate, and the enhancement of overt photorespiration and glycolate metabolism under high O_2/CO_2 environment (Tolbert, 1980). However, differences exist. Under normal environmental conditions, algae carry out a lesser amount of photorespiration presumably due to a more favorable internal low O_2/CO_2 ratio in the aqueous environment. The reduction is advantageous in the sense of reducing CO_2 loss from newly fixed carbon metabolites. Besides the reduction in the overt photorespiration, differences in the biochemical pathway also exist.

A major difference in the biochemistry of glycolate metabolism between algae and higher plants lies in the glycolate oxidizing enzyme. Some algae are similar to higher plants (tracheophytes and bryophytes) in possessing glycolate oxidase that transfers electrons directly to O_2 to form H_2O_2 (Table 4.10). Other algae contain a related enzyme, glycolate dehydrogenase, that does not transfer electrons directly to oxygen, but the native electron acceptor is unknown. The enzyme is routinely assayed using an artificial electron acceptor like 2,6-dichlorophenol indophenol, and NAD is not active. A distinct but related enzyme, D-lactate dehydrogenase, is present in many algae; its presence may confuse some investigations on the glycolate oxidizing enzymes because these latter enzymes also act on D- or L-lactate (Table 4.11). Whereas glycolate oxidase can act on L-lactate but not the D-isomer, glycolate dehydrogenase is active on D-lactate but not the L-isomer. Nevertheless, both glycolate dehydrogenase and glycolate oxidase are relatively specific for glycolate, with a K_m value 10 times lower than that for lactate. The glycolate oxidase reaction generates H_2O_2, and thus is associated with catalase so that the toxic H_2O_2 can be readily decomposed. On the other hand, the glycolate dehydrogenase reaction does not produce H_2O_2, and there is no need for the presence of catalase. Those algae possessing glycolate dehydrogenase have greatly reduced catalase activity, generally 10–50% on a protein or chlorophyll basis, than that in leaves of high plants and algae possessing glycolate oxidase (Frederick *et al.*, 1973).

In algae containing glycolate dehydrogenase, the capacity of the forward glycolate pathway is limited by the low activity of glycolate dehydrogenase which, on a basis of enzyme activity per unit of photosynthetic activity, is generally less than 10% of glycolate oxidase activity in C_3 plants (Tolbert, 1980). Excess glycolate that cannot be metabolized by the limited activity of glycolate dehydrogenase is excreted to the environment. It is estimated that up to 10% of the photosynthate in the form of glycolate is excreted to the environment under normal atmospheric CO_2 and O_2 concentrations. Since the forward glycolate pathway may not generate sufficient serine, glycine, and methylene tetrahydrofolic acid for other cellular metabolism, algae generally utilize

TABLE 4.10

Occurrence of Glycolate Oxidase and Glycolate Dehydrogenase in Algae[a]

Species	Glycolate oxidase	Glycolate dehydrogenase	Subcellular localization
Chlorophyta			
Charophyceae			
Coleochaete scutata	+		
Klebsormidium flaccidum	+		Peroxisomes (cytochemistry)
Mougeotia sp.	+		Peroxisomes (gradient isolation)
Netrium digitus	+		
Nitella sp.	+		
Spirogyra varians	+		Peroxisomes (gradient isolation)
Chlorophyceae			
Chlorogonium sp.	+		
Fasciculochloris boldii	+		
Planophila terrestris	+		
Chlorella vulgaris[a]	+	+	Glyolate oxidase in peroxisomes (gradient isolation)
Chlamydomonas reinhardtii		+	Mitochondria (cytochemistry)
Chlorella pyrenoidosa		+	
(about 20 other species)		+	
Ulvaphyceae			
Acetabularia mediterranea		+	
Codium sp.		+	
Euglenophyta			
Euglena gracilis		+	One-half mitochondria and one-half peroxisomes (gradient isolation)
Bacillariophyta			
Cylindrotheca fusiformis		+	Mitochondria (gradient isolation)
Nitzschia alba		+	
Thallassiosira pseudonana		+	
(Blue-green algae)			
Anabaena flos-aquae		+	
Oscillatoria sp.		+	

[a] The presence of dehydrogenase (Frederick *et al.*, 1973) and oxidase (Codd and Schmid, 1972) in *Chlorella vulgaris* was reported separately. The enzyme information was obtained from Frederick *et al.* (1973), Nelson and Tolbert (1970), Floyd and Salisbury (1977), and Bullock *et al.* (1979). References on subcellular localization can be found in the text. Classification of Chlorophyta follows that of Stewart and Mattox (1978).

TABLE 4.11

Comparison of Glycolate Oxidase, Glycolate Dehydrogenase, and D-Lactate Dehydrogenase[a]

	Glycolate oxidase	Glycolate dehydrogenase	D-Lactate dehydrogenase
Electron acceptor	$O_2 \rightarrow H_2O_2$	DCPIP; not NAD; may link to the mitochondrial e^- transport system	NAD; not DCPIP
Substrate specificity	Glycolate (K_m 0.38 mM)	Glycolate (K_m 0.22 mM)	
	L-Lactate (K_m 2.0 mM)	D-Lactate (K_m 1.5 mM)	D-Lactate (40 mM)
Cyanide inhibition	No	Yes[b]	
Subcellular localization	Leaf peroxisomes	Peroxisomes and mitochondria[c]	Cytosol
Occurrence	All green plants except some algae	Some algae	Many algae

[a] See text for references.
[b] On the mitochondrial enzyme only; no information on the peroxisomal enzyme.
[c] Only known in *Euglena*.

the reversible portion of the glycolate pathway (Fig. 4.6) in converting glycerate to serine and glycine more heavily than do higher plants.

2. Peroxisomes in Autotrophic Algae

Peroxisomes have been isolated successfully only from a few autotrophic algae (Tables 4.10 and 2.13), and only the results from *Euglena gracilis* have been reproduced by other workers. In most investigations, the focus has been on the subcellular localization of catalase and the glycolate oxidizing enzymes.

Among algae possessing glycolate oxidase, peroxisomes have been isolated from a few species (Table 4.10). In *Spirogyra,* the peroxisomes containing catalase, glycolate oxidase, and hydroxypyruvate reductase equilibrate at a density of 1.25 g/cm^3 on a sucrose gradient whereas the chloroplast and mitochondria equilibrated at lower densities (Stabenau, 1976). In *Mougeotia,* the peroxisomes also contain glycolate oxidase (Stabenau and Säftel, 1982). In *Chlorella vulgaris,* the isolated peroxisomes also contain catalase and glycolate oxidase (Codd and Schmid, 1972). However, there is a conflicting report claiming that glycolate dehydrogenase instead of glycolate oxidase is present in this algal species (Frederick *et al.*, 1973). In *Klebsormidium flaccidum*, cytochemistry also showed that glycolate oxidase is localized in the peroxisomes (Gruber and Frederick, 1977). From all the available information, it appears that the peroxisomes in algae that contain glycolate oxidase are similar to the leaf peroxisomes in higher plants. However, only catalase, glycolate oxidase, and hydroxypyruvate reductase are known to occur in the peroxisomes of these algae, and detailed information on other aspects is lacking.

Among those algae that contain glycolate dehydrogenase, the peroxisomes from autotrophic *Euglena gracilis* have been studied most extensively (Table 4.10). This algal species does not contain any detectable catalase activity (see Section VII,B,3). Following centrifugation of cell extracts on sucrose gradients (Fig. 4.15), the peroxisomes, mitochondria, and chloroplasts equilibrate at densities 1.25, 1.22, 1.17 g/cm^3, respectively (Collins and Merrett, 1975a,b; Yokota *et al.*, 1978a). The peroxisomes contain all the particulate hydroxypyruvate reductase and serine–glyoxylate aminotransferase, and only a portion of the particulate glutamate– glyoxylate aminotransferase, aspartate–α-ketoglutarate aminotransferase, and malate dehydrogenase. Glycolate dehydrogenase is present in about equal proportion in the peroxisomes and the mitochondria. Similar findings of dual distribution of glycolate dehydrogenase in

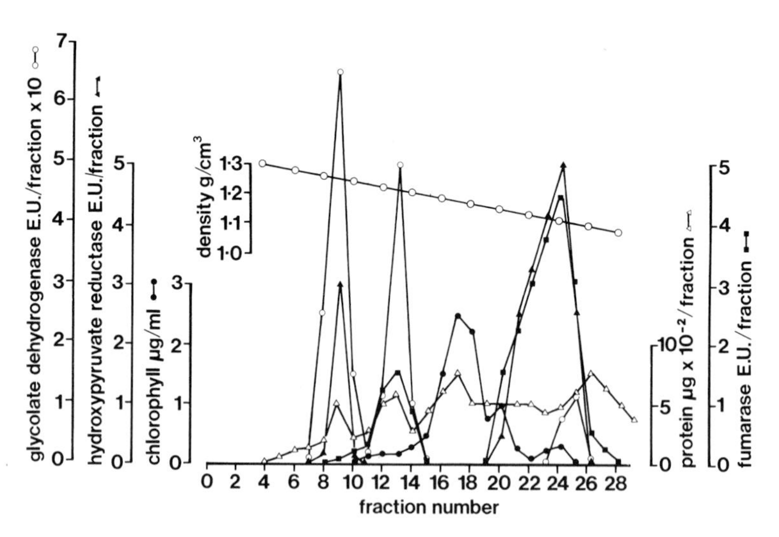

Fig. 4.15. Separation of organelles from crude cell extract of autotrophic *Euglena gracilis* in a sucrose gradient. Fraction 9 is the peak fraction of peroxisomes, and fraction 13 is the peak fraction of mitochondria. After Collins and Merrett (1975a; courtesy of Merrett).

the peroxisomes and the mitochondria in *Euglena* also were reported (Graves *et al.*, 1972; Yokota *et al.*, 1978b), although in the earlier report (Graves *et al.*, 1972) the algae were grown as heterotroph on ethanol but in the presence of periodic light. Thus, the *Euglena* peroxisomes, like the leaf peroxisomes, can metabolize glycolate to glycine. Since the peroxisomal glycolate dehydrogenase reaction does not generate H_2O_2, the absence of catalase in the peroxisomes is not deleterious. The mitochondria, with their content of glycolate dehydrogenase, glutamate–glyoxylate aminotransferase, and the glycine oxidation system, have the capacity to metabolize glycolate to glycine, serine, and CO_2. This mitochondrial capacity has been demonstrated with isolated mitochondria supplied with radioactive glycolate (Yokota *et al.*, 1978b). The mitochondrial glycolate dehydrogenase is linked to the electron transport system and eventually to O_2 as the final electron acceptor. The reversible interconversion between glycerate and serine catalyzed by hydroxypyruvate reductase and serine glyoxylate aminotransferase is restricted to the peroxisomes.

Should *Euglena* oxidize glycolate through the mitochondrial glycolate dehydrogenase, useful metabolic energy is generated through the electron transport system. In this aspect, the mitochondrial oxidation system is advantageous over the peroxisomal glycolate oxidase of higher plants and other algae. Why higher plants have evolved an apparently less efficient glycolate oxidase is a puzzle. The native electron acceptor of the algal peroxisomal glycolate dehydrogenase is unknown. The reason for *Euglena* to possess glycolate dehydrogenase in two subcellular compartments is uncertain. Whether one or both organelles oxidizes glycolate during photorespiration also is unclear.

Indirect information is available on the functional aspects of the two glycolate dehydrogenase systems in *Euglena*. Under normal conditions of illumination and atmosphere, both the peroxisomal and the mitochondrial glycolate dehydrogenase exist. Switching the growth condition to reduce photorespiration (e.g., 5% CO_2, or from autotrophy to heterotrophy) eliminates completely the peroxisomal glycolate dehydrogenase whereas the mitochondrial enzyme persists or is only partly reduced (Yokota *et al.*, 1978b). The peroxisomal enzyme, but not the mitochondrial enzyme, was thus suggested to be the one in operation during photorespiration. If the suggestion is valid, it remains to be seen how the cell would control the flow of photorespired glycolate only through the peroxisomes when the mitochondria, with their glycolate dehydrogenase and the enzymes for glycine to serine conversion, are available in the same cell.

The data on subcellular localization of glycolate dehydrogenase in *Euglena* are convincing and reproduced by other workers. Less convincing information is available on other algae possessing glycolate dehydrogenase. For two diatoms, *Cylindrotheca fusiformis* and *Nitzschia alba*, discontinuous sucrose gradients of large steps were used to isolate the peroxisomes. It was concluded that glycolate dehydrogenase was localized in the mitochondria (Paul *et al.*, 1975). However, location of the peroxisomes in the sucrose gradients was not determined (catalase and D-amino acid oxidase were undetectable), and thus the possible localization of glycolate dehydrogeanse also in peroxisomes could not be excluded. Similarly, inadequate separation of the mitochondria and peroxisomes in sucrose gradients from extracts of *Chlamydomonas reinhardtii* clouds a conclusion on whether the peroxisomes contain glycolate dehydrogenase, although the enzyme did appear to be present in the mitochondria (Stabenau, 1974). A cytochemical study on *Chlamydomonas* also supports the presence of glycolate dehydrogenase in the mitochondria (Beezley *et al.*, 1976), but again the evidence for the absence of the enzyme in the peroxisomes is not completely conclusive. The only investigation on autotrophic algae other than *Euglena* resulting in peroxisomes well separated from the mitochondria on a sucrose gradient is from *Chlorogonium elongatum;* unfortunately, only catalase and urate oxidase, but not glycolate dehydrogenase, were studied (Stabenau and Beevers, 1974).

In summary, the subcellular location of glycolate dehydrogenase and related enzymes in algae containing glycolate dehydrogenase is poorly understood. Whereas in *Euglena,* the enzyme is undoubtedly present in the mitochondria, the role of the mitochondrial enzyme in photorespiration has not been resolved. Whether algae other than *Euglena* also contain glycolate dehydrogenase in the peroxisomes is unclear. The enzymes for the interconversion between glycerate and serine are restricted to the peroxisomes in *Euglena,* but their subcellular location in other algae is unknown. With such a great taxonomical diversity in algae, one should be cautious when extending information obtained with *Euglena* to other algae without further investigations. The evolutionary relationship of the mitochondrial glycolate dehydrogenase, the peroxisomal glycolate dehydrogenase, and the peroxisomal glycolate oxidase in algae deserves further attention.

D. Peroxisomes in Ureide Metabolism

Preliminary reports (van Dijken *et al.*, 1982) indicate that cells of yeast grown on uric acid as the sole nitrogen source contain an enhanced number of peroxisomes. Presumably, these peroxisomes possess relatively high activity of urate oxidase as those in root nodule cells (Section

VI). Whether or not they contain other enzymes of the ureide pathway (Fig. 4.8) is unknown.

E. Peroxisomes in Methanol Metabolism

Methanol can be used by bacteria and some fungi as the sole carbon source for growth. As shown by radioactive tracer experiments and enzyme studies, fungal cells metabolize methanol by the following reaction:

$$\underset{\text{methanol}}{CH_3OH} \xrightarrow[\text{methanol oxidase}]{O_2 \;\; H_2O_2} \underset{\text{formaldehyde}}{HCHO} \xrightarrow{NAD \;\; NADH} \underset{\text{formate}}{HCOOH} \xrightarrow{NAD \;\; NADH} CO_2 \qquad (4.11)$$

$$HCHO \uparrow \text{(Xylulose Monophosphate Pathway)} \longrightarrow \text{assimilation}$$

Formaldehyde is either oxidized to CO_2 (for energy production) or assimilated to more complex carbon compounds by the xylulose monophosphate pathway (review by Colby *et al.*, 1979).

Methanol oxidase from *Candida boidinii* grown on methanol has been partially purified (Sahm and Wagner, 1973). It is a FAD-containing enzyme that utilizes O_2 as the electron acceptor to produce H_2O_2. It can act on methanol as well as other low molecular weight alcohols, but has the highest activity and the lowest K_m value (2.0 m*M*) on methanol. Methanol oxidase, catalase, formaldehyde dehydrogenase, and formate dehydrogenase activities are much higher in cells grown on methanol than those on other carbon sources.

The yeast methanol oxidase possesses the characteristics of a peroxisomal H_2O_2-producing flavoenzyme. The enzyme has not been reported present in the peroxisomes of higher plants. In yeast, it is localized, together with catalase, in the peroxisomes isolated from *Candida boidinii* (Fig. 4.16; Roggenkamp *et al.*, 1975), *Hansenula polymorpha* (Veenhuis *et al.*, 1978), and *Kloeckera* sp. (Tanaka *et al.*, 1976). Since the methanol oxidase reaction generates H_2O_2, its localization together with catalase in the peroxisomes is logical. Formaldehyde dehydrogenase and formate dehydrogenase are present in the cytosol (Roggenkamp *et al.*, 1975). Peroxisomes are numerous and are the dominant organelles in methanol-grown yeast (Fig. 4.17). Many of the peroxisomes contain crystalline cores with clearly visible crystalline lattice (Figs. 2.5 and 2.6).

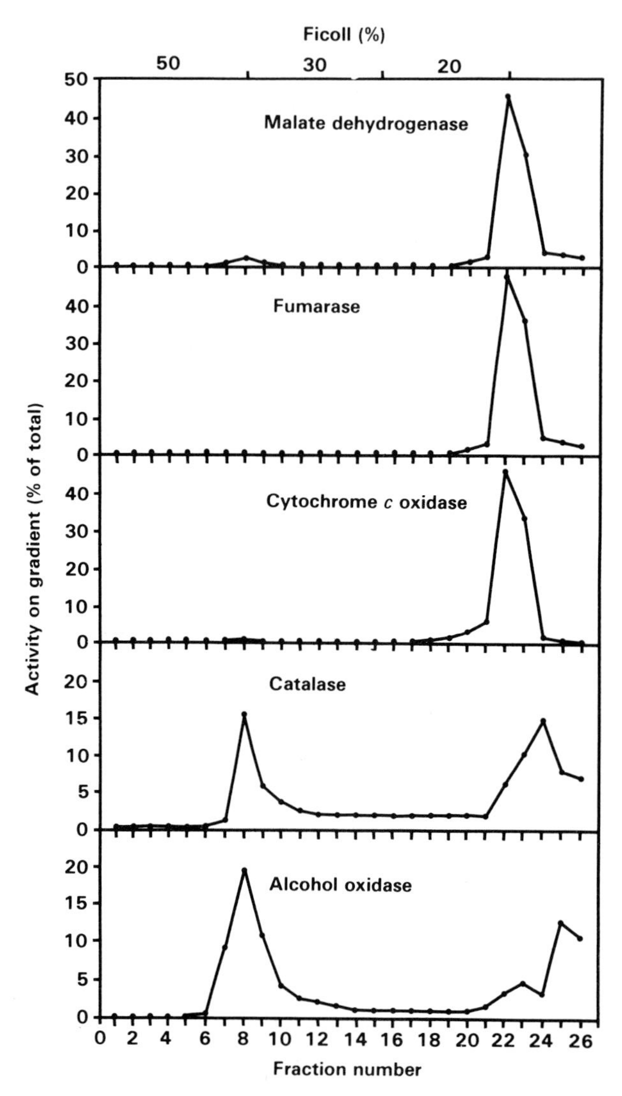

Fig. 4.16. Separation of organelles from spheroplast lysates of methanol grown *Candida boidinii* in a discontinuous Ficoll density gradient. Fumarase and cytochrome oxidase are markers of mitochondria, and catalase is a marker of peroxisomes. Redrawn from Roggenkamp *et al.* (1975).

Some bacteria utilize methanol by a serine pathway involving isocitrate lyase (Colby *et al.*, 1979). This pathway does not appear to be operative in yeast. In methanol-grown yeast cells, isocitrate lyase activity is very low, and its magnitude does not correlate well with the induction of the methanol-metabolic enzymes or the increase in number of peroxisomes (Tanaka *et al.*, 1976). It appears that methanol metabolism in yeast does not involve the glyoxylate cycle or a part of the cycle. Thus, the peroxisomes are not glyoxysome-like peroxisomes, but have a special function (Table 4.8).

F. Peroxisomes in Oxalate Synthesis

Many plant pathogenic fungi secrete oxalic acid as well as cell wall hydrolytic enzymes during infection. The process has been studied most extensively in *Sclerotium rolfsii*, a soil-borne fungus that is pathogenic to a wide range of plant species around the world. During infection, the fungus produces and secretes oxalic acid and pectinases. The oxalic acid

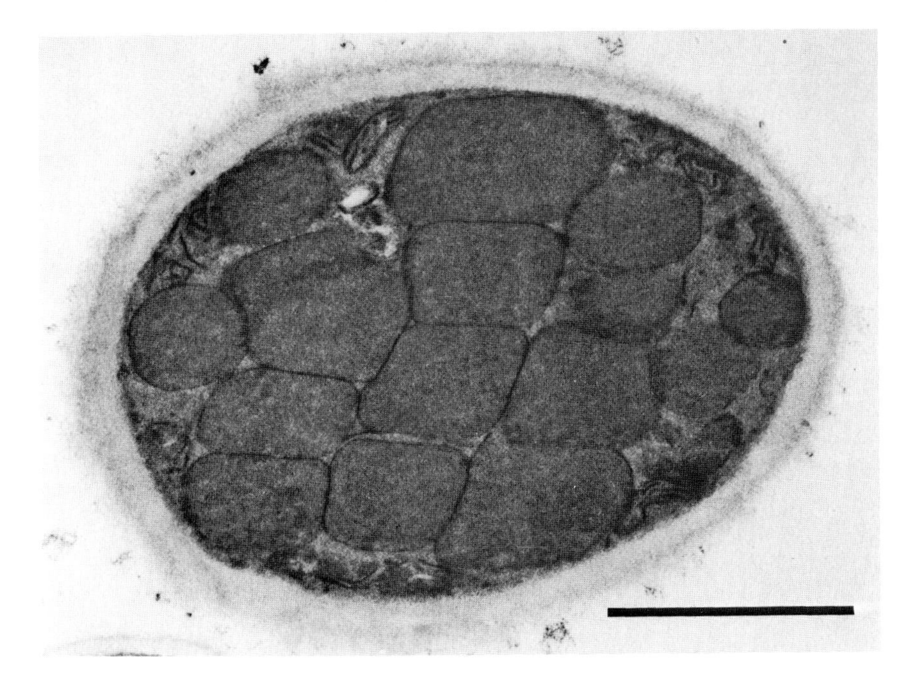

Fig. 4.17. Section of a *Hansenula polymorpha* cell grown on methanol as the sole carbon source, showing numerous peroxisomes. Bar = 0.5 μm. From Veenhuis *et al.* (1978).

creates a lower pH environment for the acidic pectinases, and at the same time loosens the cell wall by chelating the cell wall calcium (Bateman and Beer, 1965). Oxalic acid production also occurs when the fungus is grown on extracted host cell wall polysaccharides or other artificial media, but is greatly reduced when grown on a simple carbon source at a low pH or unbuffered media.

Oxalic acid is produced by the following enzymatic reactions:

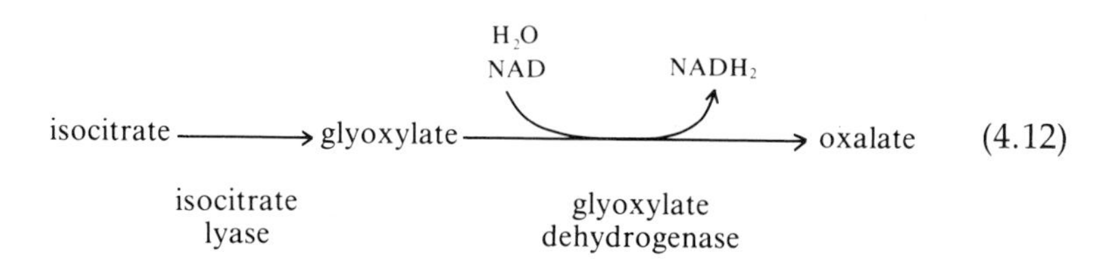

The properties of NAD-glyoxylate dehydrogenase were studied in a cell-free extract of *S. rolfsii* (Maxwell and Bateman, 1968). The enzyme utilizes NAD, but not NADP, O_2, or 2,6-dichlorophenol indophenol as the electron acceptor. The K_m for glyoxylate is 1.2×10^{-4} *M*, and the enzyme does not act on glycolate, ethanol, lactate, or pyruvate at 1.3 m*M* or higher concentrations. The specificity of the enzyme toward the substrate and electron acceptor suggests that it is not lactate dehydrogenase (Sawaki *et al.*, 1967), xanthine oxidase (Gibbs and Watts, 1966), or glycolate oxidase (Richardson and Tolbert, 1961) that can oxidize glyoxylate to oxalate. This enzyme has not been described in other organisms.

During the active infection period, numerous peroxisomes are present in the fungal hypha (Table 2.8). The two enzymes, isocitrate lyase and glyoxylate dehydrogenase, together with catalase, are localized in peroxisomes (Fig. 4.18). No malate synthase was detected in the peroxisomes, suggesting that isocitrate lyase does not participate in the glyoxylate bypass. Whether isocitrate is derived from the Krebs cycle in the mitochondria or the peroxisomes contain their own citrate synthase and aconitase for isocitrate production is unknown.

The two key enzymes, isocitrate lyase and glyoxylate dehydrogenase, are not involved in H_2O_2 production. However, their localization in the peroxisomes is in accord with the idea that the peroxisomes compartmentalize reactions involving glyoxylate so that this reactive metabolite cannot be diverted into other undesirable reactions. Although spinach and a few other plant species produce and accumulate substantial amounts of oxalate, glyoxylate dehydrogenase was not found in isolated spinach leaf peroxisomes (Chang and Huang, 1981).

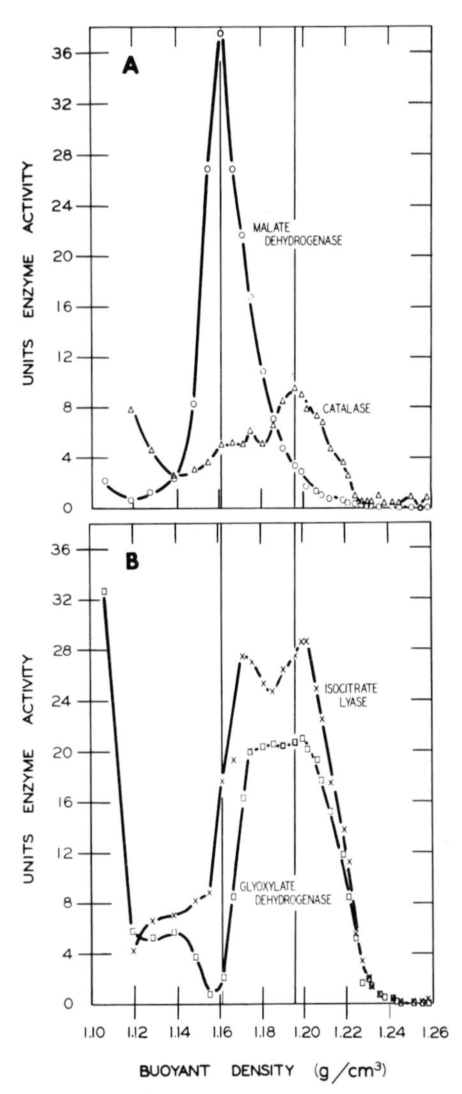

Fig. 4.18. Separation of organelles from a 3,000 *g* supernatant from *Sclerotium rolfsii* homogenate on a linear sucrose gradient in a zonal rotor. Vertical lines mark densities at which peak activities of malate dehydrogenase (mitochondrial marker) and catalase (peroxisomal marker) occur. From Armentrout *et al.* (1978).

G. Peroxisomes in Amine Metabolism

Yeasts of many genera can use various amines as the sole nitrogen source for growth (Yamada *et al.*, 1965; Zwart *et al.*, 1980; van Dijken and Bos, 1981). These amines include aliphatic mono-, di-, and triamines of short carbon chains, and simple aromatic amines.

Monoamines are metabolized by the following reaction:

$$RCH_2NH_2 + H_2O \xrightarrow[\text{amine oxidase}]{O_2 \quad H_2O_2} RCHO + NH_3 \qquad (4.13)$$

Ammonia is assimlated to organic nitrogen compounds, whereas RCHO is oxidized further in the presence of formaldehyde dehydrogenase. In general, yeasts cannot utilize amines as the sole sources of both nitrogen and carbon, suggesting that they do not assimilate RCHO. The above amine oxidation may be involved in the catabolism of ethanolamine and choline from phospholipids. Yeasts can grow on ethanolamine or choline as the sole nitrogen source. Ethanolamine is an active substrate for amine oxidase. Choline is converted to trimethylamine and subsequently to monomethylamine, which is then oxidized to formaldehyde in the presence of amine oxidase.

Amine oxidase from *Aspergillus niger* is a copper-containing enzyme that is most active toward aliphatic monoamines of C_3 to C_6 (Yamada *et al.*, 1965). Amine oxidase, together with catalase, is localized in the peroxisomes in *Candida utilis* and *Hansenula polymorpha* (Fig. 4.19) (Zwart *et al.*, 1980). The two enzymes and formaldehyde dehydrogenase are induced after the yeasts have been transferred to media containing amines as the sole nitrogen source. Under this growth condition, the number of peroxisomes also is enhanced (Fig. 4.20).

Several yeast species are able to utilize two or even three different classes of nitrogen compounds (amines, urate, and D-amino acids) via hydrogen peroxide producing oxidases. In addition, some yeast species can grow on amines (or the other two classes of nitrogen compounds) as the nitrogen source and methanol or alkanes as the carbon source. However, yeast species capable of utilizing both methanol and alkanes as the carbon source have not been found. When *Hansenula polymorpha* cells are grown in media containing methylamine as the nitrogen source *and* methanol as the carbon source, the activities of both amine oxidase and methanol oxidase are present in each individual peroxisomes (Veenhuis *et al.*, 1982).

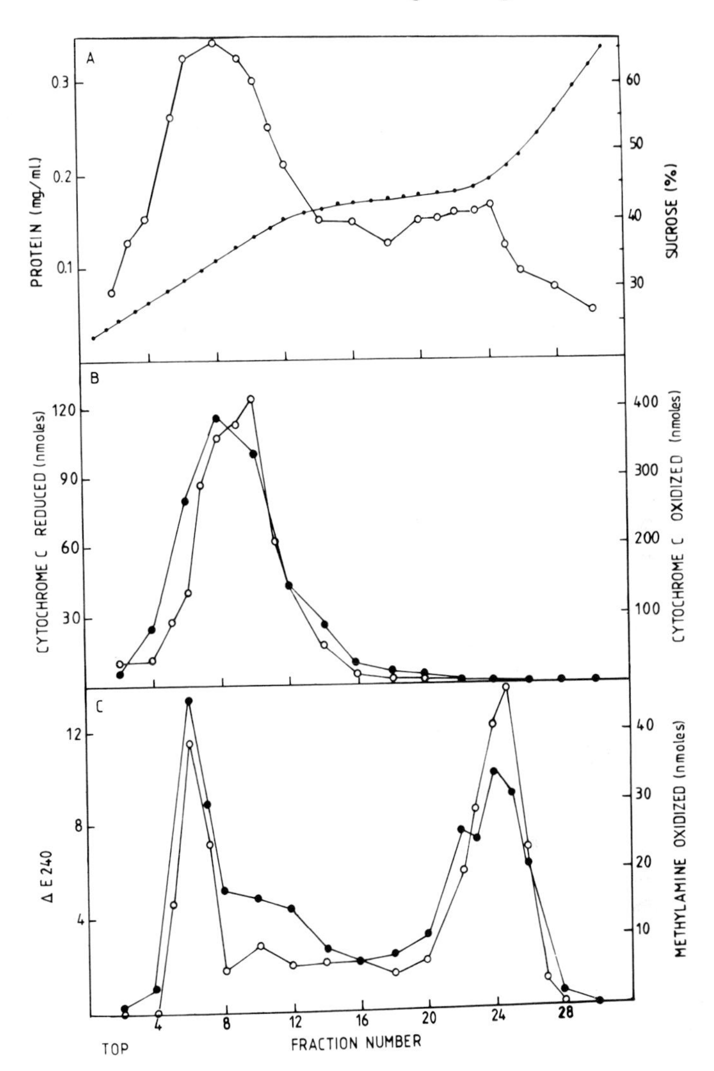

Fig. 4.19. Separation of organelles on a discontinuous sucrose gradient from a particulate fraction (organelles sedimented at 20,000 *g* from a supernatant obtained after centrifugation of the homogenate at 14,000 *g*) of extract from *Candida utilis* grown on methylamine as the nitrogen source. NADH-cytochrome reductase (●, membrane marker) and cytochrome oxidase (○, mitochondrial marker) in Part B, and catalase (ΔE_{240}, ●, peroxisomal marker) and methylamine oxidase (○) in Part C are shown. Enzyme activities are expressed in nmoles/min/ml. The peroxisomal enzyme activities at the top of the gradient were due to organelle breakage during resuspension of the particulate fraction. Courtesy of K. Zwart (unpublished).

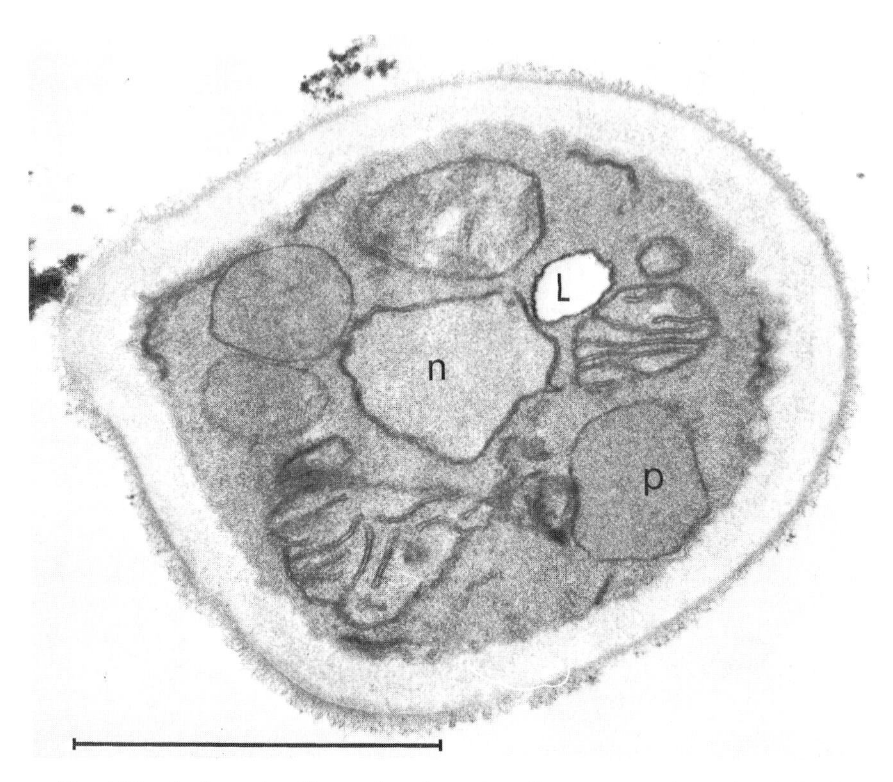

Fig. 4.20. Section of a *Hansenula polymorpha* cell grown on glucose/methylamine, showing peroxisomes (p), nucleus (n), lipoid droplet (L), and double-membraned mitochondria. The cell was fixed with potassium permanganate. Bar = 1 μm. From Zwart *et al.* (1980).

VIII. OTHER ENZYMES REPORTED PRESENT IN PEROXISOMES

The enzymes present in the various types of peroxisomes of higher plants, fungi, and algae are summarized in Table 4.12. A few additional enzymes also have been reported present in the peroxisomes of various plant tissues; these claims have not been supported by convincing data or confirmed by other laboratories. In some instances, the enzyme activity present in the peroxisomes represents such a small fraction of the total cellular activity that the possibility of contamination has to be eliminated before the claim can be fully accepted. It is by no means to say that these enzymes are not localized in the peroxisomes; simply, further evidence is needed to justify the claims.

Nitrate reductase, nitrite reductase, and enzymes of the reductive pentose phosphate pathway were reported to be "loosely associated" with the leaf peroxisomes of tobacco, and this association was affected by light and hormonal treatment (Lips, 1975). The current belief (Beevers and Hagemen, 1980) is that, in leaves and roots, nitrate reductase is a cytosolic enzyme and nitrite reductase is associated with plastids. Enzymes for the metabolism of aromatic acids were reported to be present in the leaf peroxisomes and glyoxysomes, but the recovery of the total enzyme activities in the peroxisomes was less than 10% (Kindl and Ruis, 1971). Shikimate dehydrogenase was reported in the peroxisomes of pea leaves (Rothe, 1974), but subsequently it was suggested that this finding was due to plastid contamination (Feierabend and Brassel, 1977).

Superoxide dismutase is involved in the metabolism of H_2O_2 in all aerobic organisms (Fridovich, 1975; McCord, 1979). The enzyme catalyzes the following reaction:

$$O_2^- + O_2^- + 2\,H^+ \rightarrow H_2O_2 + O_2 \qquad (4.14)$$

The enzyme presumably protects the cells from oxygen toxicity. The highly reactive superoxide generated enzymatically or nonenzymatically under high oxygen tension is converted to H_2O_2 by superoxide dismutase. Although the enzymatic reaction produces H_2O_2, the enzyme has never been found in plant or animal peroxisomes where catalase is localized. Perhaps the H_2O_2 produced is destroyed by peroxidase outside the peroxisomes.

Several species of *Aspergillus* and *Penicillium* secrete an antibiotic that was initially called notatin (after *P. notatum*). Subsequently, this antibiotic was identified as glucose oxidase which catalyzes the following reaction:

$$\underset{\text{glucose}}{C_6H_{12}O_6} + O_2 + H_2O \rightarrow \underset{\text{gluconic acid}}{C_6H_{12}O_7} + H_2O_2$$

The antibiotic effect is dependent on the presence of glucose and oxygen and is due to the product H_2O_2 rather than gluconic acid or the acidity (Coulthard *et al.*, 1945). The enzyme is used extensively in industry for food protection (e.g., removing oxygen or desugaring). It has a molecular weight of 154,000 (*Penicillium* enzyme) or 186,000 (*Aspergillus* enzyme). The *Penicillium* enzyme is composed of two FAD-containing subunits of molecular weight 81,000 each of which contains two polypeptides of molecular weight 45,000 (Yoshimura and Isemura, 1971). The activity is highest at pH 5.6 to 5.8, and the apparent K_m value for glucose is 0.015 M (Bentley, 1963). Although the *Penicillium* enzyme is routinely isolated from the culture medium, the *Aspergillus* enzyme is apparently

TABLE 4.12

Enzymes in Peroxisomes of Higher Plants, Fungi, and Algae[a]

	Higher plants				Fungi and algae								
	1	2	3	4	1	2		3		4	5	6	7
	Potato tuber	Castor bean endosperm	Spinach leaf	Soybean nodule	Spirogyra on glucose	Candida on alkane	Euglena on acetate	Spirogyra, autotrophic	Euglena, autotrophic	Hansenula on urate	Candida on methanol	Sclerotium, pathogenic	Candida on amines
Catalase	+	+	+	+	+	+	−	+	−	+	+	+	+
Glycolate oxidase	+	+	+		+	+		+					
Xanthine dehydrogenase (oxidase)		−		−									
Urate oxidase	+	+	+	+		+				+			
Allantoinase		+		−									
D-Amino acid oxidase	−					+							
Hydroxypyruvate reductase		+	+					+	+				
NADP-isocitrate dehydrogenase			+			+							
NAD-cytochrome reductase		+	+										
Alkaline lipase		+											
Fatty alcohol dehydrogenase						+							
Fatty aldehyde dehydrogenase						+							
Fatty acyl-CoA synthetase		+				+							
β-Oxidation enzymes			+[b]			+							
Fatty acyl-CoA oxidase		+											

Enoyl-CoA hydratase	+							
β-Hydroxyacyl-CoA dehydrogenase	+							
Thiolase	+							
Acetyl-CoA synthetase	+		+	+				
Isocitrate lyase	+		+	+			+	
Malate synthase	+		+	+			−	
Malate dehydrogenase	+	+	−	+	+			
Aconitase	+		−	+				
Citrate synthase	+		−	+				
Aspartate–α-ketoglutarate aminotransferase	+	+			+			
Serine–glyoxylate aminotransferase		+			+			
Glutamate–glyoxylate aminotransferase		+			+			
Special function enzymes								
Methanol oxidase						+		
Glycolate dehydrogenase		−			+			
Glyoxylate dehydrogenase		−					+	
Creatine acetyltransferase			+					
NAD-glycerol-phosphate dehydrogenase			+					
Amine oxidase								+

[a] The numbers underneath higher plants, fungi, and algae refer to the types of peroxisomes categorized according to their physiological roles, as listed in Table 4.8. The peroxisomes from one to two tissues/organisms are selected to represent each type of peroxisome. The selection of these tissues/organisms is based on choosing the best-studied system within each type of peroxisome (potato, castor bean, spinach, soybean, *Candida* on alkane, *Euglena*-autotrophic, *Candida* on methanol, and *Sclerotium*-pathogenic), or the uniqueness of the organisms (*Euglena* on acetate containing the complete glyoxylate cycle enzymes in the peroxisomes; autotrophic *Spirogyra* representing those algae with glycolate oxidase), or the availability of information (*Spirogyra* on glucose).

[b] Low activities, see Table 4.1.

also present intracellularly. The activities of glucose oxidase and catalase in cell-free extract of *A. niger* are higher when the fungus is grown in aerated fermenter than in shake flasks. Cytochemistry shows that these two enzymes are present in the peroxisomes (van Dijken and Veenhuis, 1980). The physiological role of the peroxisomal glucose oxidase is unknown, as is its relationship with the extracellular enzyme.

IX. SUMMARY

Peroxisomes in various tissues contain as basic enzymatic constituents catalase and one or more H_2O_2-producing oxidases. The organelles in some tissues possess additional enzymes and are important cellular components with unique metabolic roles. In higher plants, peroxisomes can be catagorized into four types according to their possible physiological roles:

1. The "unspecialized" peroxisomes are minor cellular components occurring in a great variety of tissues. They contain only the basic peroxisomal enzymes and play no known physiological role, and therefore are tentatively described as "unspecialized."

2. The glyoxysomes are specialized peroxisomes in the storage tissues of oilseeds. They contain β-oxidation and glyoxylate cycle enzymes, and function in the conversion of fatty acid to succinate, which is used for gluconeogenesis or synthesis of other biosynthetic intermediates.

3. The leaf peroxisomes containing active enzymes of the glycolate pathway are specialized peroxisomes in photosynthetic tissues. They carry out the major reactions of photorespiration.

4. The peroxisomes in the uninfected root nodule cells of certain legumes contain high urate oxidase activity. They synthesize allantoin, which is the major metabolite for nitrogen transport within these plants.

In fungi and algae, the peroxisomes generally contain catalase and one or more H_2O_2-producing oxidase. The organelles may possess certain additional enzymes, unknown in higher plant peroxisomes, to carry out special metabolic roles. These specialized peroxisomes often enlarge and increase in number when the microorganisms are grown under special trophic conditions. Seven types of fungal/algal peroxisomes have been identified:

1. The "unspecialized" peroxisomes (analogous to the term used for higher plants) occur in heterotrophic fungi and algae grown on glucose, lactose, or sucrose.

2. The glyoxysome-like peroxisomes exist in fungi and algae grown on

alkane, fatty acid, ethanol, and acetate. They generally possess only two of the five glyoxylate cycle enzymes.

3. The peroxisomes in autotrophic algae contain either glycolate oxidase or glycolate dehydrogenase, and participate in photorespiration.

4. The peroxisomes in fungi grown on urate as the sole nitrogen source contain high activity of urate oxidase for nitrogen metabolism.

5. The peroxisomes in fungi grown on methanol contain methanol oxidase and catalase. The organelles oxidize methanol to formaldehyde, which will be either oxidized further to CO_2 (for energy production) or assimilated to complex carbon compounds.

6. The peroxisomes in phytopathogenic fungi contain isocitrate lyase and glyoxylate dehydrogenase. The organelles produce oxalic acid, which is secreted onto the host tissues to aid infection.

7. The peroxisomes in fungi grown on amines contain amine oxidase and catalase. The organelles oxidize amines to ammonia which is assimilated.

It is anticipated that peroxisomes with other metabolic roles will be identified in the future, such as those in microorganisms grown on glycolate or D-amino acid as the sole carbon and/or nitrogen source. Little morphological and biochemical information is available on the peroxisomes in bryophytic tissues, and red and brown algae, and their similarities with the peroxisomes in tracheophytic tissues, fungi, and algae are unknown.

5

Ontogeny

I. INTRODUCTION

The development of form and function of peroxisomes has received dramatically increased attention in recent years. Such studies are important in understanding the control of functioning of the peroxisomes alone, and in relationship with other organelles. In addition, the study of peroxisome biogenesis is an excellent opportunity for investigating the flow of membranes and proteins among different compartments of a cell. Thus far, the known biogenesis mechanism of peroxisomes appear to be distinct from those of other organelles. In spite of the apparent structural simplicity of the peroxisome, the synthesis and assembly of the organelle appear to be somewhat complicated. Peroxisomes do not contain their own DNA (Chapter 3), and the synthesis of the organelle must be controlled elsewhere in the cell, presumably the nucleus.

Historically, the observations that peroxisomes in rat liver have a close connection with the endoplasmic reticulum (ER) and apparently receive catalase precursors from the rough ER (Kashiwagi *et al.*, 1971, but unverified) have led to the suggestion that peroxisomes are produced by the vesiculation of a specialized region of the ER. This ER-vesiculation model is in accordance with the well-documented mechanism of segregating of secretory proteins into secretory vesicles in mammalian tissues. Indeed, secretory vesicles in excretory cells have morphological features quite similar to those of peroxisomes, namely, an amorphous protein matrix surrounded by a single membrane. Studies on glyoxysome biogenesis in tissues of germinated oilseeds have been directed toward following the ER-vesiculation model, and much circumstantial evidence obtained earlier has been interpreted to reinforce this concept. However, new data from studies of both oilseed glyoxysomes and rat liver peroxisomes have provided evidence that is inconsistent with many aspects of the ER-vesiculation model and modified models are being formulated.

Most of the studies on peroxisome biogenesis have centered on the glyoxysomes in oilseeds. Studies on the biogenesis of leaf peroxisomes generally have limited to the appearance of leaf-type peroxisomes in greening cotyledons concomitant with the disappearance of glyoxysomes in germinated seeds. Studies on the formation of peroxisomes during leaf development are few. Although the biogenesis of peroxisomes (mostly glyoxysomes) has been investigated quite extensively, little effort has been devoted toward understanding degradation of the organelles.

II. THE ENDOPLASMIC RETICULUM-VESICULATION MODEL

This model of peroxisome biogenesis follows the model for synthesis and sequestration of secretory proteins in mammalian tissues as outlined by Palade (1975) and detailed further by Blobel's group (Blobel and Dobberstein, 1975 and several subsequent papers). The latter workers used the term "signal hypothesis" to describe their model. This hypothesis states that for a certain protein, particularly one to be inserted into and through a membrane, a precursor form exists which has an extra amino acid sequence on the amino-terminal (first-synthesized) end. This sequence "signals" that insertion into a membrane is to occur. The hypothesis is illustrated in Fig. 5.1, which shows a protein being synthesized on ribosomes bound to the ER and cotranslationally inserted into the lumen. The "signal" sequence on the NH_2 terminus inserts into the ER immediately upon synthesis and the rest of the polypeptide follows. In the lumen, this signal sequence is removed by a specific protease. For certain glycoproteins, the proteins have been shown to be glycosylated

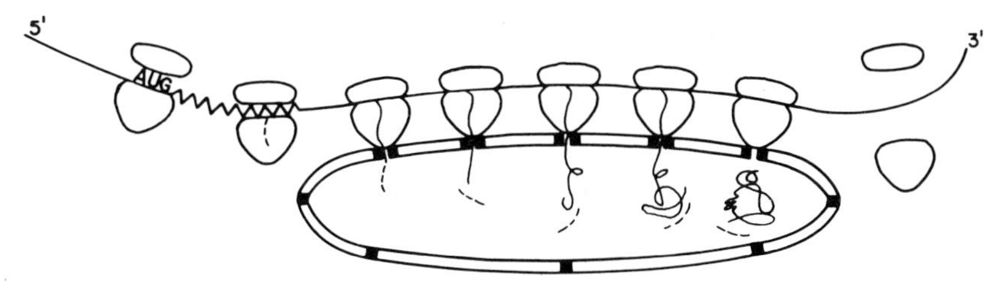

Fig. 5.1. Schematic representation of the signal hypothesis (Blobel and Dobberstein, 1975). See text for explanation.

in the ER. For nonsecretory proteins functional in the cytosol, the signal sequence might not exist or it might be different, so that insertion into the ER lumen during or after synthesis on free polysomes would not occur.

Until recently, peroxisomes with their constituent proteins were thought to be formed by vesiculation of specialized regions of the ER. The peroxisomal matrix proteins presumably were synthesized on polysomes bound to ER, and cotranslationally inserted and modified, and segregated within the lumen of the ER. Then, the proteins would move to a specialized region of the ER which would then pinch off to become a new peroxisome (Fig. 5.2). Support for this ER-vesiculation model of peroxisome biogenesis rested on several lines of observations. Although some of these observations have been subsequently demonstrated to

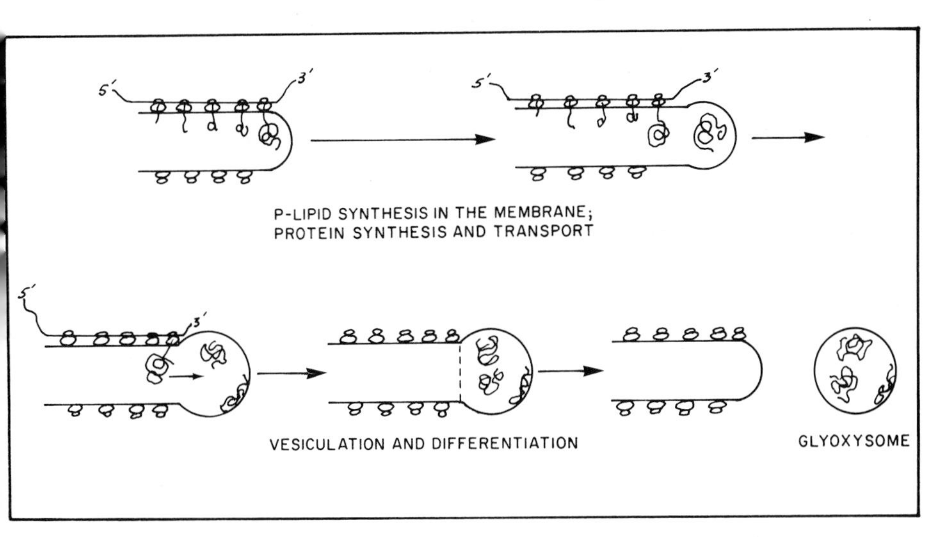

Fig. 5.2. Proposed scheme for biogenesis of glyoxysomes in oil seeds from segments of rough endoplasmic reticulum. Modified from Beevers (1979).

have arisen from artifacts, other observations are still maintained valid but the interpretation has been greatly modified.

The observations used to support the ER-vesiculation model are summarized briefly in this section; comments on their validity will be discussed in greater detail together with recent findings in the next section. These observations are as follows.

1. Electron microscopic observations indicate a close association between the peroxisomes and the ER. However, a direct connection between the lumen of the ER and the matrix of the peroxisome, with a continuous membrane occurring between the two organelles has been observed rarely (e.g., in cotton cotyledons during embryogenesis; Fig. 5.3), and has not been reported in any tissues of germinated oil seeds.

2. The peroxisomes do not have the capacity to synthesize membrane phospholipids, and most phospholipid synthesis occurs in the ER. This observation still holds true.

3. The phospholipids (by thin-layer chromatography) and proteins (by sodium dodecyl sulfate–acrylamide gel electrophoresis) of the ER and the glyoxysomal membranes have some similar components. These observations have been disputed or the interpretation questioned.

4. In castor bean endosperm, both malate synthase and citrate synthase occur in the glyoxysomal and ER fractions isolated in sucrose density gradients. The enzymes in each fraction appear to have identical

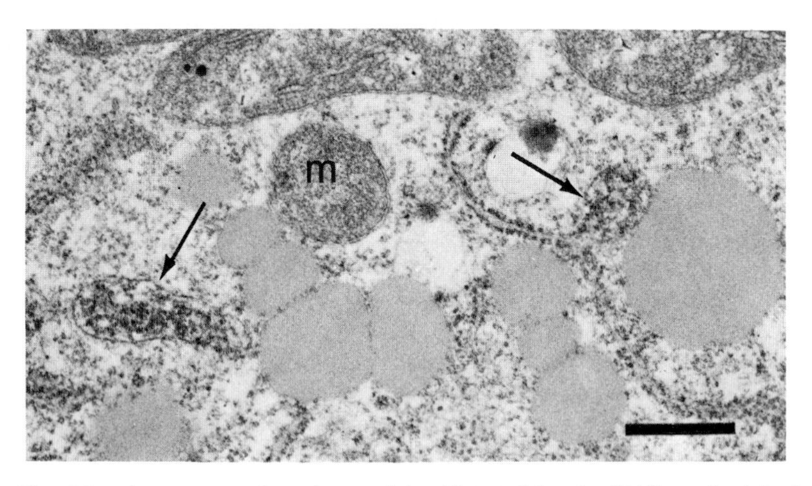

Fig. 5.3. Appearance of catalase activity (detected by the DAB reaction) in two smooth-surfaced vesicles connected to rough endoplasmic reticulum (arrows) in a cotyledon cell of a cottonseed embryo. The embryo was harvested 22 days after anthesis (see Fig. 5.11 for enzyme development during cottonseed embryogenesis) m, mitochondrion. Bar = 0.5 μm. From Choinski and Trelease (1978).

chemical properties and immunological homologies. In addition, at early stage of postgerminative growth, concomitant with the increase in total glyoxysomal enzyme activities, relatively more malate synthase and citrate synthase are observed in the ER fraction than at a later stage of seedling growth when enzyme activities no longer are increasing rapidly. This observation has been questioned in view of the existence of aggregated enzymes contaminating the ER fraction.

5. *In vivo* pulse-chase labeling of phospholipids, proteins, and glycoproteins with, respectively, radioactive choline, methionine, and mannose indicates the appearance of newly synthesized phospholipids or proteins in the ER and subsequently chased to the glyoxysomes.

6. Castor bean glyoxysomal membranes contain activities of several tightly bound enzymes that are also present in the ER fraction (Section IV,F, Chapter 4). These enzymes include alkaline lipase, cytochrome b_5, cinnamic acid 4-hydroxylase, *p*-chloro-*N*-methylaniline-*N*-demethylase, NADH-cytochrome *c* reductase, and NADH-ferricyanide reductase.

The ER-vesiculation model in a highly modified version may still hold true (next section), although it would be distinctly different from the ideas originally set forth.

III. PEROXISOME BIOGENESIS*

The validity of the ER-vesiculation model gradually loses support in view of new experimental observations. Although the phospholipids of the peroxisomal membrane still are believed to be synthesized in the ER, all the individual proteins examined so far appear to be synthesized in the cytosol rather than on the rough ER. The mode of assembling individual protein and membrane phospholipid components is still unknown.

A. Biosynthesis of Individual Components

1. Membrane Lipid

The ER plays a central role in the synthesis of peroxisomal phospholipids. In castor bean endosperm, the bulk of the phospholipid synthesis occurs in the ER (Moore, 1982). The evidence comes from results obtained by measuring activities of enzymes of phospholipid synthesis in various organelles separated in sucrose density gradients. Figure 5.4 gives representative data for results of two phospholipid synthesizing

*As of late-1982.

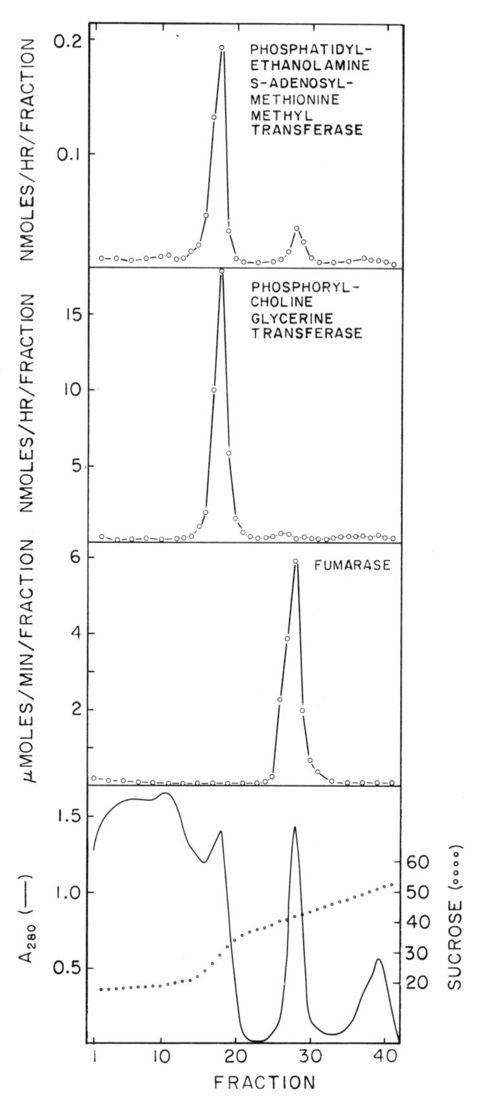

Fig. 5.4. Intracellular localization of lecithin synthesis by methylation of phosphatidylethanolamine and by phosphorylcholine glyceride transfer. Both activities are predominantly in the endoplasmic reticulum, but the methylation also occurs in the mitochondria. From Moore (1976).

enzymes, showing a major contribution by the ER and some activities associated with the mitochondria. A summary of the findings from castor bean is presented in Table 5.1. Similar results have been summarized for spinach leaves (Kates and Marshall, 1975). No activity of phospholipid-synthesizing enzymes has been found in the peroxisomal fraction of any tissues, leading to the conclusion that peroxisomes are incapable of synthesizing their own membrane lipids.

In castor bean seedlings, the newly synthesized phospholipids appeared first in the ER and then was transferred to the glyoxysomes (Lord, 1978). The experimental results obtained by feeding radioactive choline to the tissue of different ages and then assessing the amount of radioactive choline incorporated into the lipids of isolated organelle fractions are shown in Fig. 5.5. One-day-old tissue contains radioactivity only in the ER and none in other organelles. By day three, radioactivity still occurs in the ER, but increasing proportions are present in other organelles, including the glyoxysomes. Thus, the formation of ER phospholipid precedes their incorporation into peroxisomal membranes. Furthermore, pulse-chase experiments utilizing [^{14}C]choline and 4-day-old castor bean endosperm (a period of active glyoxysome synthesis) demonstrate that the incorporated choline appears first in the ER of the cells and then, as it is lost from the ER, increases in the mitochondria and glyoxysomes (Fig. 5.6). Thus, phospholipids destined for the glyoxy-

TABLE 5.1

Compartmentation of Phospholipid Biogenesis in Castor Bean Endosperm[a]

	Percentage of total activity				
Lipid formed	Soluble fraction	Microsomes	Mitochondria	Plastids	Glyoxysomes
Fatty acids	0	0		100	0
Phosphatidate	0	87	13	0	0
Diacylglycerol	Present	Present	?	?	0
Phosphatidylcholine	0	94	6	0	0
Phosphatidylethanolamine	0	98	2	0	0
Phosphatidylserine	0	100	0	0	0
CDP-diacylglycerol	0	75	25	0	0
Phosphatidylglycerol	0	50	50	0	0
Bisphosphatidylglycerol	?[b]	?	Present	?	?
Phosphatidylinositol	0	100	0	0	0

[a] From Moore (1982).
[b] ?, undetermined.

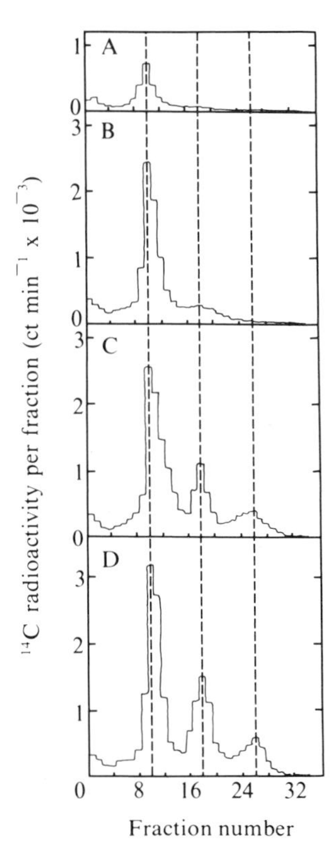

Fig. 5.5. Distribution of radioactive phosphatidylcholine in organelles isolated from imbibed castor bean (A), and endosperm of 1-day-old (B), 2-day-old (C) and 3-day-old (D) castor bean seedlings. The tissue had been incubated with radioactive choline for 3 hr. Marker enzyme activities for endoplasmic reticulum, mitochondria, and glyoxysomes peaked in fractions 10, 18, and 26, respectively. Each gradient represents the organelles recovered from homogenizing 10 endosperm halves. Redrawn from Lord (1978).

somes appear to be produced in the ER and then transferred to the glyoxysomes.

A similarity exists between the phospholipids and fatty acids of the ER and glyoxysomal membranes of castor bean endosperm (Table 5.2). This similarity was interpreted as support for the two membranes arising in contact with each other according to the ER-vesiculation model (Donaldson and Beevers, 1977). However, subsequent evidence obtained by measuring the susceptibility of phospholipids in the membranes of the intact organelles to phospholipase A_2 attack casts doubt on this interpretation (Cheesbrough and Moore, 1980). Although the phospholipid compositions of the two types of organelles are similar, the actual distribution of the phosphatidylcholine between the inner and outer leaflets of the two organelle membranes is different (Table 5.3).

It seems beyond doubt that the peroxisomal membrane lipids are synthesized in the ER and somehow become incorporated into the peroxisomal membrane lipids. The possible mechanisms of assembly of phospholipids with proteins to form the peroxisomes will be discussed later.

2. Membrane Proteins

a. Integral Proteins. When glyoxysomal membranes prepared by osmotic lysis of isolated glyoxysomes are treated with 0.15 *M* KCl solution, the peripheral proteins including malate synthase and citrate synthase are solubilized, whereas the integral proteins including alkaline lipase and several electron-transport proteins and presumably also structural proteins remain in the ghosts (Chapter 3).

It has been shown earlier that in castor bean endosperm, the peripheral proteins of the ER membrane and glyoxysomal membrane share

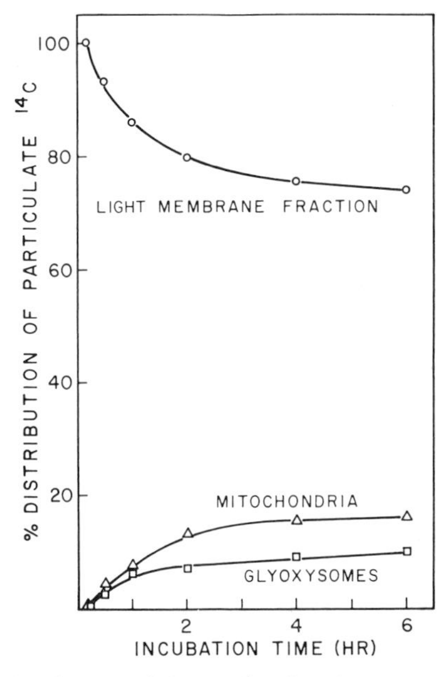

Fig. 5.6. Percent distribution of the total radioactivity among the endoplasmic reticulum (light membrane fraction), mitochondria, and glyoxysomes varying with time after application of [^{14}C]choline to detached endosperm of 4-day-old castor bean seedlings. From Kagawa *et al.* (1973a).

TABLE 5.2

Phospholipid Content (Estimated by Phosphorus Content) of Various Organelles Isolated from Castor Bean Endosperm[a]

	Phospholipid content (mole %)[b]			
	Glyoxysome	Endoplasmic reticulum	Mitochondria	Proplastid
Phosphatidylcholine	51.4	45.3	33.1	35.6
Phosphatidylethanolamine	27.2	28.7	35.7	38.7
Phosphatidylinositol	9.0	13.1	3.9	8.0
Phosphatidylserine	1.5	2.3	2.2	4.6
Phosphatidylglycerol[c]	2.7	3.6	2.2	8.4
Cardiolipin	2.3	2.5	17.7	—
Origin	2.4	1.5	1.6	2.7
Front	1.5	1.0	1.7	1.7
Other	2.0	2.0	0.9	0.3
Number of experiments	4	3	4	2

[a] From Donaldson and Beevers (1977).
[b] Recovery of phosphorus from thin layer chromatography was 87% or better.
[c] Includes phosphatidic acid.

many similar components as shown by sodium dodecyl sulfate–acrylamide gel electrophoresis (Bowden and Lord, 1976a,b); these results were interpreted as evidence of biogenetic relationship. Subsequently, nonanalogy of the membrane proteins was reported (Moreau *et al.*, 1980; Kindl, 1982). Also, earlier results suggesting that some integral (and peripheral also) proteins were glycosylated (Bowden and Lord, 1976a) have been disputed (Bergner and Tanner, 1981; Lord and Roberts, 1982), and there is no firm evidence for the existence of glycosylated proteins in peroxisomes (Chapter 3, Section IV).

The mode of synthesis and insertion of integral membrane proteins is totally unknown. These proteins have been studied only in terms of their solubility and molecular weights (Chapter 3). In view of their hydrophobic, transmembrane, and possibly glycosylated nature, they may be synthesized on rough ER, perhaps in the vicinity of the newly formed phospholipids to generate new membrane complex. Pulse-chase experiments (Lord and Roberts, 1982) utilizing radioactive methionine or mannose and 4-day-old castor bean endosperm demonstrate that the incorporated methionine or mannose appears first in the KCl-washed ER membrane and then, as it is lost from the ER, increases in the KCl-washed glyoxysomal membrane (Fig. 5.7). Thus, the ER may be involved in the synthesis of at least some glyoxysomal integral membrane proteins.

*b. **Peripheral Proteins.*** The studies of the biosynthesis of membrane peripheral proteins (this section) and matrix proteins (next section) carried out in several laboratories have utilized similar techniques. First, the tissue is labeled with radioactive [^{35}S]methionine and the enzyme in tissue extracts is immunoprecipitated with antibodies raised previously against the purified enzyme. The molecular weight of the radioactive enzyme is then assessed by sodium dodecyl sulfate–acrylamide gel electrophoresis followed by radioautography/fluorography. In most studies, the *in vivo* labeled enzyme has a subunit molecular weight similar to that of the enzyme isolated by traditional methods. Second, the poly(A)-mRNA is isolated from the tissue and allowed to direct an *in vitro* protein synthesis in a complete protein-synthesizing system with radioactive amino acids. Afterward, the enzyme is immunoprecipitated and analyzed by electrophoresis. The molecular weight of *in vivo* and *in vitro* synthesized enzymes are compared. This second procedure is repeated using poly(A)-mRNA derived from

TABLE 5.3

Summary of Phospholipid Distributions in Mitochondrial, Glyoxysomal, and Endoplasmic Reticulum Membranes[a]

		Distribution (mole %)			
		Mitochondrial membranes			
Lipid class	Location	Outer	Inner	Glyoxysomal membrane	Endoplasmic reticulum
Phosphatidylcholine	Outer leaflet	25	21	22	52
	Inner leaflet	30	20	78	20
	Inaccessible	45	59	0	28
Phosphatidyl-ethanolamine	Outer leaflet	44	70	67	65
	Inner leaflet	32	18	16	15
	Inaccessible	24	12	17	20
Phosphatidylserine, plus phosphatidylinosital	Outer leaflet	50	0	7	0
	Inner leaflet	25	100	93	0
	Inaccessible	25	0	0	100
Total	Outer leaflet	32	53	51	43
	Inner leaflet	37	10	45	28
	Inaccessible	31	37	4	29

[a] Compiled from Cheesbrough and Moore (1980 and unpublished data).

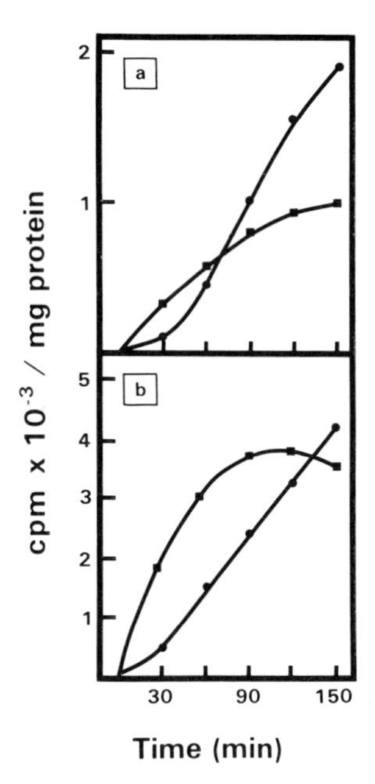

Fig. 5.7. Kinetics of labeling of membrane proteins and glycoproteins. Intact endosperm halves were incubated with (a) [^{35}S]methionine or (b) [^{14}C]mannose. At various times, the tissue was homogenized. KCl-washed membranes were prepared and their radioactivity determined. ■, endoplasmic reticulum membranes; ●, glyoxysomal membranes. Redrawn from Lord and Roberts (1982).

either the cytosol or the rough ER as a means to assess the subcellular site of synthesis of the enzyme. Third, the tissue is labeled with radioactive [^{35}S]methionine, and then its extract subfractionated into fractions of peroxisomes, ER, cytosol, and other organelles. The enzyme in each of the fractions is immunoprecipitated, and the time course of labeling and the pulse-chase labeling in each subcellular compartments is assessed. A summary of the findings is shown in Table 5.4.

i. Malate Synthase. Malate synthase and citrate synthase in castor bean are two peripheral membrane proteins that can be solubilized by 0.15 *M* KCl (Chapter 3). The changing distribution of these two enzymes in organelles fractions isolated from castor bean endosperm following germination was studied (Gonzalez and Beevers, 1976; Kagawa and Gonzalez, 1981). In two-day-old seedlings, a high proportion of malate synthase is associated with the ER fraction; by day four the relative activity in this fraction is considerably reduced and that in the glyoxysomal fraction is increased. Citrate synthase behaves similarly. Catalase has a higher proportion of activity in the soluble fractions at the earlier

TABLE 5.4

The Subunits and Their Molecular Weights of Glyoxysomal and Leaf-Peroxisomal Enzymes Synthesized *in Vivo* and *in Vitro* by poly(A)-mRNA Extracted from the Tissues

Enzyme	Subunit composition	Tissue	Molecular weight		Intracellular compartment where the newly synthesized subunits first appear	Synthesized on free/bound polysomes or 70 S/80 S ribosomes	References
			in vivo	*in vitro*			
Malate synthase	Octomer	Cucumber cotyledon	57,000	57,000	—	—	Becker *et al.* (1982)
		Cucumber cotyledon	63,000	63,000	Cytosol	—	Köller and Kindl (1980)
		Castor bean endosperm	(same)		Cytosol	—	Lord and Roberts (1982)
Isocitrate lyase	Tetramer	Cucumber cotyledon	63,000 61,500	61,500 60,000	—	—	Becker *et al.* (1982)
		Cucumber cotyledon	64,000	64,000	Cytosol	—	Frevert *et al.* (1980)
		Castor bean endosperm	(same)		Cytosol	Free polysomes	Lord and Roberts (1982)
		Neurospora	(same)		—	Free polysomes	Zimmermann and Neupert (1980)
Multifunctional protein	Monomer	Cucumber cotyledon	75,000	75,000	Cytosol	—	Frevert *et al.* (1980)
Glycolate oxidase	Dimer	*Lens* leaves	43,000	43,000	—	—	Kindl (1982)
Catalase	Tetramer	Cucumber cotyledon	53,500	57,000	Cytosol	—	Kindl *et al.* (1980)
		Cucumber cotyledon	54,000	55,000	—	—	Becker *et al.* (1982)
Malate dehydrogenase	Dimer	Watermelon cotyledon	33,000	38,000	—	80 S ribosomes	Walk and Hock (1978)
		Cucumber cotyledon	33,000	38,000	—	—	Becker *et al.* (1982)

stages of development, and does not occur in the ER fraction at any stage. The catalytic and immunological properties of the two synthases in the ER fraction are similar to those of the glyoxysomal enzymes (and for citrate synthase, unlike those of the mitochondrial isozyme). Thus, it was suggested that malate synthase and citrate synthase are first synthesized in the ER and then transferred to the budding glyoxysomes.

Strong evidence indicates that this occurrence of malate synthase, and perhaps citrate synthase, in the ER fraction is only fortuitous (Köller and Kindl, 1980; Kindl, 1982). When the ER fraction of castor bean extract was pooled and the organelles were allowed to either sediment or float on another sucrose density gradient, most of the malate synthase was not associated with the ER markers but behaved as large protein aggregates. Furthermore, in cucumber cotyledon, three kinds of malate synthase molecules recognized by immunoprecipitation can be found *in vivo:* 5 S (monomer, molecular weight 63,000), 19 S (octomer, native enzyme), and 100 S (large aggregate, which sedimented with the ER fraction in sucrose gradients). These three species can be converted to one another *in vitro* by manipulating the salt and hydrophobic environment. It is still being argued that in castor bean, most (Gonzalez, 1982) or about half (Lord and Roberts, 1982) of the malate synthase in the ER fraction is indeed associated with the ER marker.

It is perhaps of little importance as to whether or not malate synthase in the ER fraction is indeed intimately associated with the ER. Malate synthase, being a peripheral membrane protein, should have a binding affinity with the glyoxysomal membrane; if it could somehow adsorb to the ER artificially, some enzymes lost from glyoxysomes during cell breakage and fractionation would be expected to contaminate the ER fraction even after purification and washing. The more important question is whether the malate synthase in the ER fraction is newly synthesized enzyme. This has been studied in cucumber cotyledons (Koller and Kindl, 1980). Poly(A)-mRNA directed *in vitro* synthesis of malate synthase yields the 5 S species. *In vivo* labeling of protein with [^{35}S]methionine reveals that the 5 S and 19 S species in the cytosol are labeled first with the highest specific radioactivity, and subsequently, the 19 S in the glyoxysomes is labeled (Fig. 5.8). The 100 S (the "ER" enzyme) always has a low specific radioactivity, and thus should not be the new enzyme to be incorporated into the glyoxysomes.

In cucumber cotyledon (Kindl, 1982; Becker *et al.*, 1982) and castor bean endosperm (Lord and Roberts, 1982), poly(A)-mRNA directed *in vitro* synthesis of malate synthase yields subunits (the 5 S species) of the same molecular weight as the subunit of enzyme isolated from the tissues. Apparently, there is no covalent cleavage of the polypeptide

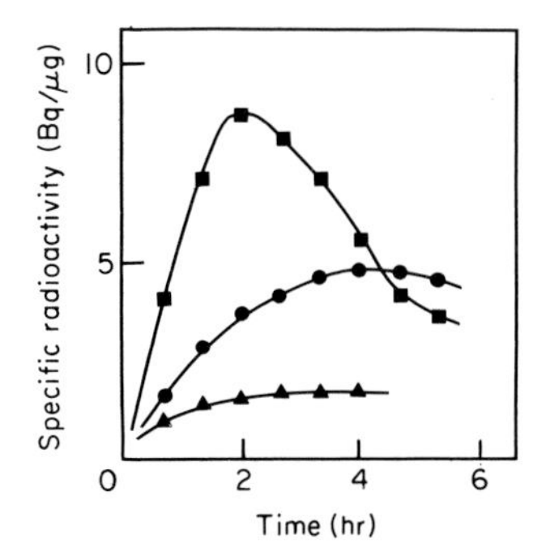

Fig. 5.8. Change of specific radioactivities of three malate synthase pools in cotyledons of 3-day-old cucumber seedlings. The tissue was supplied with [^{35}S]methionine at time 0. The specific radioactivities of the cytosolic pool (5 S + 19 S, upper curve), glyoxysomal pool (19 S, middle curve), and "microsomal" malate synthase (100 S, lower curve) are shown. Redrawn from Kindl (1982).

cotranslationally or posttranslationally. The addition of a mammalian microsomal preparation (containing cotranslationally cleaving protease) does not result in any polypeptide cleavage of the malate synthase subunit, even though as a control it does cleave the newly synthesized castor bean phytoagglutinin (a protein known to be modified cotranslationally). Earlier work in support of the ER-vesiculation model suggested that malate synthase was a glycoprotein and thus might be synthesized on rough ER and that the newly synthesized polypeptides were glycosylated, and perhaps otherwise modified, during its segregation into the lumen of the ER. It is now clear, however, that malate synthase in castor bean is not a glycoprotein (Bergner and Tanner, 1981; Lord and Roberts, 1982).

Malate synthase exhibits amphipathic binding properties. The 19 S species is bound to the glyoxysomal membrane by ionic interaction, since 0.15 *M* KCl can strip the enzyme off the membrane (Chapter 3). In cucumber, the three malate synthase species bind to externally added phospholipids, and the 5 S species also binds to lipid membranes (Kindl, 1982). Furthermore, the 5 S species newly synthesized from poly(A)-mRNA *in vitro* associates with intact glyoxysomes such that the incorporation renders the enzyme resistant to protease action.

In summary, the newly synthesized malate synthase appears first in the cytosol and, presumably but not necessarily, is synthesized on free polysomes. The nascent enzyme is similar to the glyoxysomal enzyme in subunit molecular weigth. It is not a glycoprotein. There is no indication of a cotranslational or posttranslational modification of the enzyme.

ii. Malate Dehydrogenase. Malate dehydrogenase is only loosely associated with the glyoxysomal membrane, and can be removed by 0.05 *M* KCl (Chapter 3). In watermelon cotyledons (Walk and Hock, 1978), poly(A)-mRNA directed *in vitro* synthesis of malate dehydrogenase produces four protein bands of different molecular weights (Fig. 5.9). The major band, which is the largest, has a molecular weight of 38,000. Of the three minor bands, the smallest one has a molecular weight of 33,000, which is similar to the subunit molecular weight of purified glyoxysomal malate dehydrogenase. *In vivo* pulse-chase experiments with [^{35}S]methionine indicate that the large subunit of molecular weight 38,000 is the precursor form of the small subunit of molecular weight

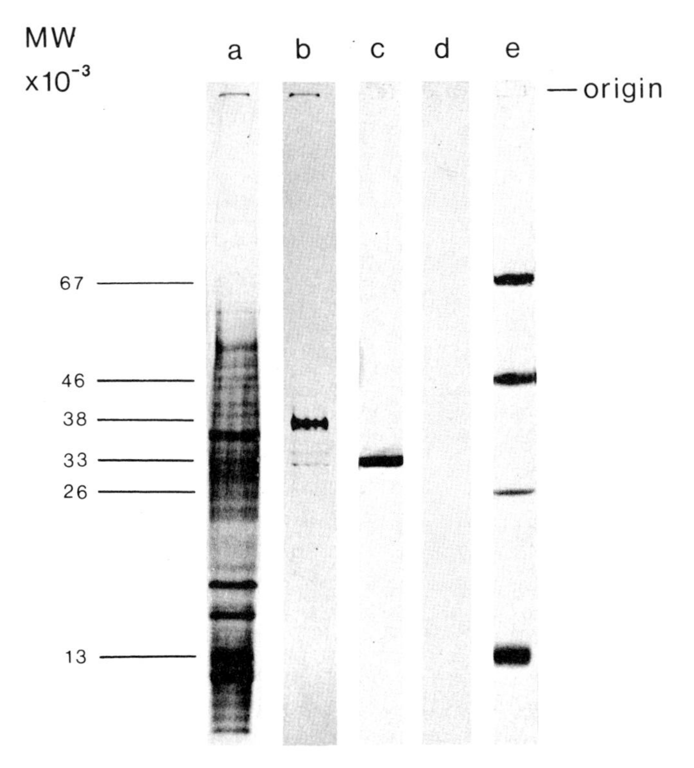

Fig. 5.9. Evidence for a high molecular weight precursor of glyoxysomal malate dehydrogenase. Sodium dodecyl sulfate–polyacrylamide gel electrophoresis patterns were determined for polypeptides synthesized *in vitro* by watermelon polyadenylated mRNA. Lanes c and e were stained for protein, the remainder were visualized by autoradiography. (a) Total radioactive protein synthesized; (b) radioactive protein precipitated by antibodies against malate dehydrogenase; (c) purified glyoxysomal malate dehydrogenase; (d) protein precipitated by antibodies from the rabbit before induced immunity to the enzyme; (e) molecular weight standards: cytochrome *c* (13,000), chymotrypsinogen (26,000), ovalbumin (16,000), and bovine serum albumin (67,000). From Walk and Hock (1978).

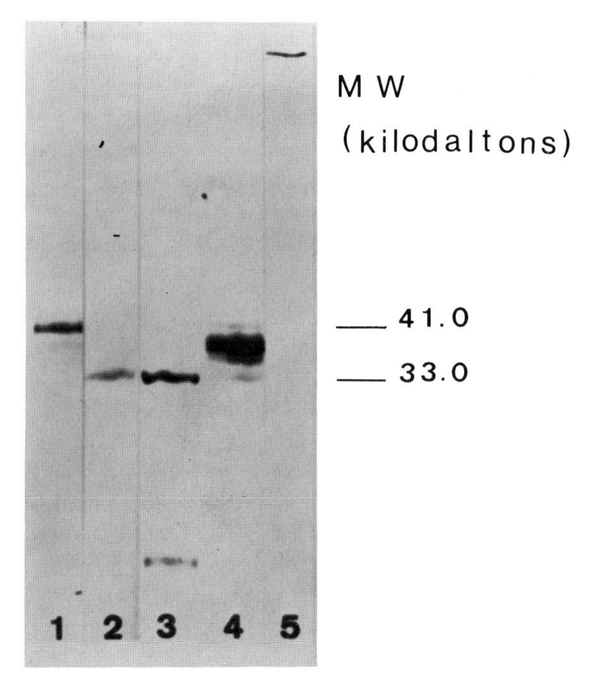

Fig. 5.10. *In vitro* processing of glyoxysomal malate dehydrogenase. Fluorogram of sodium dodecyl sulfate–polyacrylamide gel electrophoresis patterns of polypeptides synthesized *in vitro* by watermelon polyadenylated mRNA is shown. Lane 1, *in vitro* synthesized glyoxysomal malate dehydrogenase; 2, *in vivo* synthesized glyoxysomal malate dehydrogenase as immunoprecipitated from crude extract of tissue pretreated with [^{35}S]methionone; 3, *in vitro* transport of glyoxysomal malate dehydrogenase [After *in vitro* synthesis, the complete reaction medium was incubated with a crude organelle fraction (10,000 *g* pellet) for 1 hr at 25°C. After treated with proteinase K and terminated the treatment by phenymethanesulfonyl fluoride, the organelles were shocked and centrifuged. Glyoxysomal malate dehydrogenase was immunoprecipitated from the supernatant.]; 4, same as lane 3 but without protease treatment; 5, same as lane 3 but with protease treatment after shocking the organelles. From Hock and Gietl (1982).

33,000. Similar results were obtained from cucumber cotyledons (Becker *et al.*, 1982). In watermelon (Hock and Gietl, 1982), incubation of the newly synthesized subunit of 38,000 with a crude organelle fraction (containing glyoxysomes, ER, and other organelles) leads to an incomplete processing of the subunit (i.e., fragments of subunits of molecular weight between 38,000 and 33,000 are produced), whereas a subsequent addition of protease to the incubation mixture generates a complete processing of the large subunit into the small subunit (Fig. 5.10). One possible interpretation of these results is that the large subunit is processed in the cytosol by cytosolic protease before its entry to the glyoxysome. After the *in vitro* association of the enzyme with the organelles, presumably glyoxysomes, externally added protease can still recognize and hydrolyze the extra polypeptide sequence. Thus, malate dehydrogenase would have to be localized on the outer surface of the organelle. While this latter interpretation is interesting, it is possible that the observed association of the newly synthesized malate dehydrogenase with the glyoxysomes (or with the ER, or the mitochondria) may be artificial under the incubation conditions; malate dehydrogenase as a peripheral membrane enzyme should have an affinity for membrane structures.

iii. Multifunction Protein (Enoyl-CoA Hydratase and β-Hydroxyacyl-CoA Dehydrogenase). Similar to malate dehydrogenase, the multifunctional protein of the β-oxidation sequence is only loosely associated with the glyoxysomal membrane and can be solubilized by 0.05 *M* KCl (Chapter 3). Its biosynthesis has been studied only in cucumber cotyledon (Kindl, 1982). The enzyme isolated from the cotyledon has a molecular weight of 75,000 (no subunit), as is the enzyme synthesized *in vitro* by cotyledon poly(A)-mRNA. There is no apparent cotranslational or posttranslational modification of the protein.

3. Matrix Proteins

a. Catalase. The biogenesis of catalase in plants has been studied only in cucumber cotyledon, and the information is still unclear. According to one report (Becker *et al.*, 1982), catalase isolated from the tissue has a subunit molecular weight of 54,000, whereas the catalase synthesized *in vitro* by poly(A)-mRNA has a subunit molecular weight of 55,000. According to another report (Kindl, 1982), the *in vivo* enzyme has a subunit molecular weight of 53,000, and the catalase synthesized by cotyledon poly(A)-mRNA *in vitro* has a subunit molecular weight of 57,000. Thus, it is possible that there is a sequence that is cleaved when the enzyme subunit passes through a membrane into the matrix of either a budding vesicle of the ER or an enlarging peroxisome. However, the *in vitro* product of higher molecular weight is resistant to digestion by added protease (Kindl, 1982). It should be noted that the *in vitro* product does not contain heme (molecular weight approximately 600), and the binding of heme to the polypeptide could change the structure of the subunit such that its molecular weight on sodium dodecyl sulfate gels would be altered. The question of whether a "signal" sequence is present in the nascent catalase subunit should remain open. The extra sequence, if it exists, may be cleaved in the cytosol before or during the entry of the subunit into the ER vesicle or peroxisome.

In rat liver, catalase is synthesized only on free polysomes (Goldman and Blobel, 1978). The nascent catalase subunit is present first in the cytosol and then in the peroxisomes. The enzyme molecule appears first as subunits in the peroxisome, and subsequently four subunits join together to form the tetrameric enzyme (Redman *et al.*, 1972; Lazarow, 1981). It is still unknown whether the heme is added to each of the subunits before or after the entry of the subunit to the peroxisomes. Clearly, the ER is not involved in catalase biosynthesis. Whether this mode of assembly also occurs in plant peroxisome remains to be seen.

b. Isocitrate Lyase. With cucumber cotyledons, poly(A)-mRNA directed *in vitro* synthesis of isocitrate lyase produces subunits of molecular weight 64,000, the same as the molecular weight of *in vivo* synthesized enzyme subunits (Kindl, 1982). A report from a different laboratory indicates that the *in vitro* products contain protein subunits of molecular weights of 61,500 and 60,000, which are slightly smaller than the molecular weight of 63,000 of the enzyme subunits isolated from the tissue (Becker *et al.*, 1982). In assessing the above data, one should note that isocitrate lyase is an enzyme that can be subjected easily to artificial modification (Theimer, 1976; Khan *et al.*, 1977; Frevert *et al.*, 1980). With castor bean endosperm, poly(A)-mRNA directed *in vitro* synthesis also produces enzyme subunits with a molecular weight the same as that of the enzyme subunit isolated from that tissue (Lord and Roberts, 1982). Furthermore, the enzyme is synthesized on free polysomes. *In vivo* pulse-chase radioactive labeling with [^{35}S]methionine shows that the newly formed enzyme is present first in the cytosol and then appears in the glyoxysomes (Fig. 5.11). At no time is a significant proportion of the total isocitrate lyase associated with the ER. Apparently, the ER is not involved in either the synthesis of isocitrate lyase or its assembly into the newly formed glyoxysomes.

In *Neurospora* grown on acetate, isocitrate lyase also is synthesized by free polysomes (Zimmermann and Neupert, 1980). Poly(A)-mRNA directed *in vitro* synthesis of isocitrate lyase produces a subunit of molecular weight the same as the subunit of the authentic enzyme. The newly synthesized enzyme is incorporated into a crude organelle fraction (presumably the glyoxysomes); this incorporation renders the enzyme insusceptible to protease digestion.

c. Glycolate Oxidase. Glycolate oxidase is only a minor enzyme constituent in glyoxysomes but is a major one in leaf peroxisomes. With *Lens* leaves, the poly(A)-mRNA directed *in vitro* synthesis of glycolate oxidase produces enzyme subunits of molecular weight 43,000, exactly

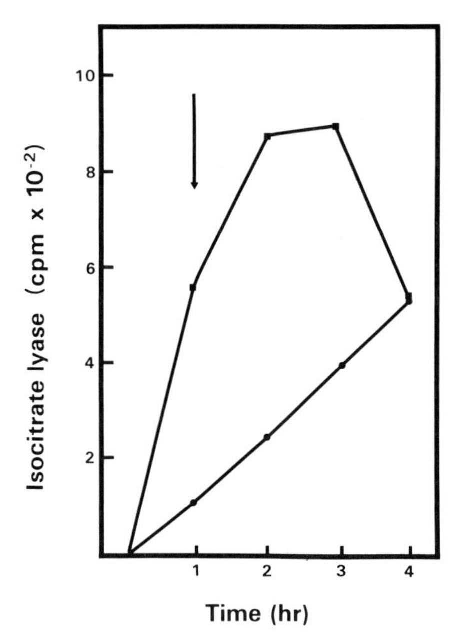

Fig. 5.11. Pulse-chase labeling of cytosolic and glyoxysomal isocitrate lyase in castor bean endosperm. The tissue was incubated with [^{35}S]methionine for 1 hr, and 1000-fold excess of unlabeled methionine was administered (indicated by arrow). At the times indicated, six halves of endosperm were homogenized and fractionated. Isocitrate lyase was immunoprecipitated from cytosol (■) and glyoxysomal (●) fractions, separated electrophoretically, and the radioactivity content of the protein band assayed. From Lord and Roberts (1982).

the same as those isolated from the tissue (Kindl, 1982). Apparently, a signal sequence is absent from the subunit.

d. Urate Oxidase. The biogenesis of urate oxidase has not been studied in plant tissues. In rat liver, this enzyme, together with catalase, is synthesized on free polysomes (Goldman and Blobel, 1978).

B. Temporal Development of Peroxisomes

According to the ER-vesiculation model of peroxisome biosynthesis, the organelle and its enzymatic content are synthesized at approximately the same time so that a totally new organelle is produced. This latter concept may or may not hold true in view of the latest experimental results on peroxisome biosynthesis.

In ungerminated castor bean, the activities of malate synthase, and perhaps other glyoxysomal enzymes also, are almost completely absent.

This situation may be unique to castor bean. In other oilseeds examined, malate synthase activity is present in the ungerminated stage (Miernyk *et al.*, 1979). In cotton cotyledons during seed development (Choinski and Trelease, 1978; Miernyk and Trelease, 1981a,b), activities of the β-oxidation enzymes, catalase, and malate synthase, but not isocitrate lyase, increase and reach a level that is 15–19% of the peak activities in postgermination (Fig. 5.12). These enzyme activities are present in peroxisomes and persist through seed desiccation. Following mature seed imbibition, these activities increase again (severalfold), and are accompanied by a rapid increase of isocitrate lyase activity from a zero level in dry seeds to a peak at about the same time as the other enzymes. At the onset of enzyme development (about 22 days after anthesis) (Fig. 5.12), evaginations from segments of rough ER possessing catalase reactivity in embryonic cells can be observed by electron microscopy (Fig. 5.3). These evaginations were interpreted as showing the origin of peroxisomes which proliferate and gain enzymes throughout maturation (Miernyk and Trelease, 1981a). A similar, but not identical, situation has been described for cucumber cotyledons (Kindl, 1982). At an early stage of seed maturation, catalase and multifunctional protein (two β-oxidation enzymes) are present first in the cytosol and subsequently in the peroxisomes which lack malate synthase and isocitrate lyase. At the late stage of cucumber seed maturation, substantial malate synthase activity in addition to low levels of isocitrate lyase activity are present in the cucumber cotyledons. The endosperm in developing castor beans also is capable of fatty acid β-oxidation (Hutton and Stumpf, 1969).

The results of the above findings can be interpreted in several ways. The peroxisomes proliferated during seed maturation may carry out a unique metabolic function specific for normal seed maturation (Chapter 4, Section IV,G). Another possible interpretation is that these peroxisomes are "incomplete organelles" produced due to gene leakage. The most attractive interpretation is that the embryo-synthesized peroxisomes may be progenitors of the glyoxysomes that function during postgerminative growth. Implicit in this interpretation is the necessity for addition of newly synthesized enzymes (and perhaps of membrane components also) into the preexisting organelles. This would be in accord with current data showing posttranslational insertion of enzymes into glyoxysomes during postgerminative growth, but in contrast to the ER-vesiculation model. Since more than 80% of all the glyoxysomal enzymes still have to be synthesized in postgermination (Fig. 5.12), it is necessary to determine if these enzymes are present in organelles newly synthesized in postgermination or they are added to preexisting organelles synthesized during seed maturation. To provide some basic

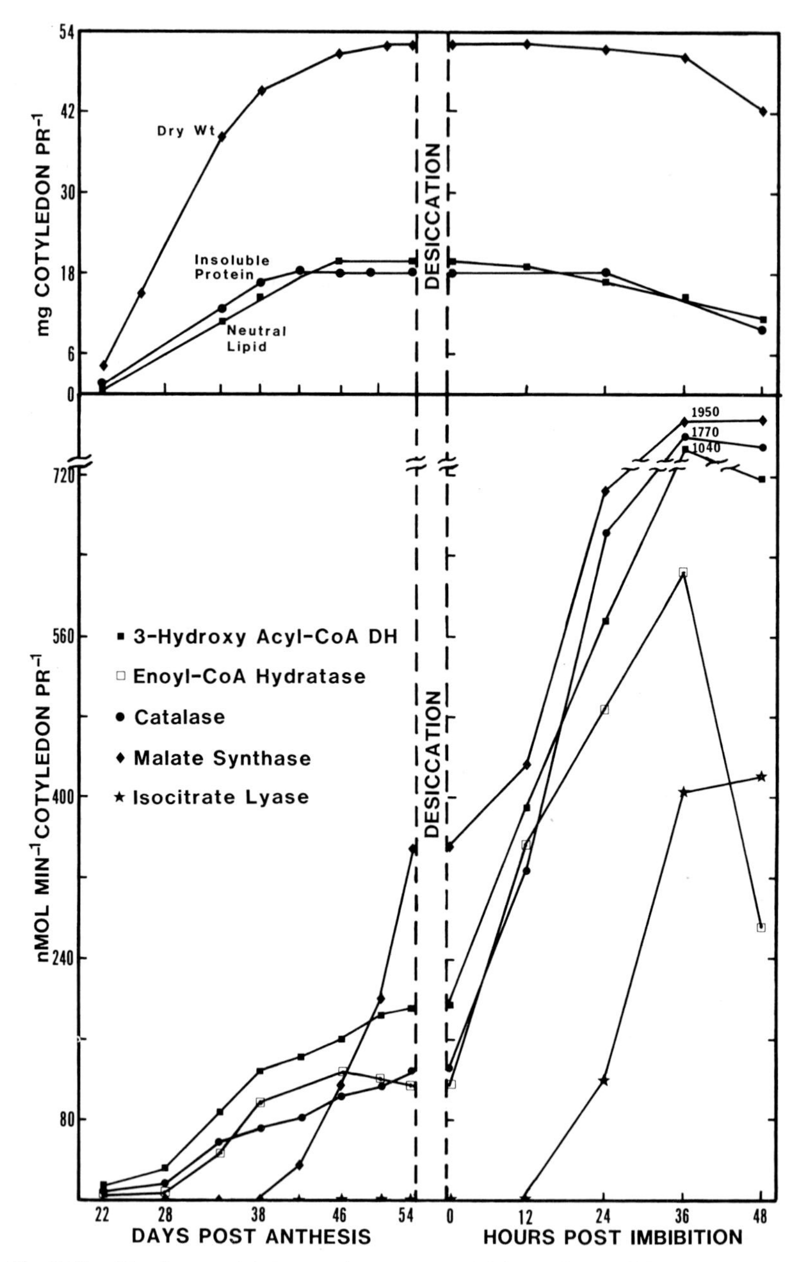

Fig. 5.12. Developmental changes in enzyme activities and neutral lipid content in the cotyledons of cotton seed during maturation and germination. The relative enzyme activities per cotyledon pair are shown. From R. N. Trelease (unpublished graph).

information, the number of peroxisomes per cell during late seed maturation and postgermination should be determined.

C. Integration of Information*

1. Current Knowledge

The current information can be summarized into the following:

1. The phospholipids of the peroxisomal membrane are synthesized by the ER.

2. The site of membrane integral protein synthesis is unknown, but the ER is a likely candidate.

3. Of all the membrane peripheral proteins and matrix proteins that have been examined, none has been shown to be synthesized on rough ER. In contrast, many of them have been shown to have their nascent molecules appear first in the cytosol or be synthesized by free polysomes. Malate synthase and the multifunctional protein (membrane peripheral proteins), as well as isocitrate lyase and glycolate oxidase (matrix proteins), are synthesized initially as subunits of molecular weights identical to those of the authentic enzymes in the peroxisomes. Malate dehydrogenase (membrane peripheral protein) is synthesized as a subunit (MW 38,000) larger than the subunit (MW 33,000) of the final enzyme. Catalase (matrix protein) may be synthesized initially as subunits slightly larger than those of the authentic enzyme (MW 55,000 vs. 54,000), but the incorporation of heme group to the protein may affect the observed molecular weights.

4. There has been only limited success in the *in vitro* incorporation of newly synthesized glyoxysomal enzymes into glyoxysomes.

These findings can be interpreted to fit to one of, but not limited to, the biogenesis models illustrated in Fig. 5.13. The possible mechanisms of phospholipid and protein assembly are discussed in Sections III,C,2 and III,C,3.

2. Assembly of Phospholipids

Phospholipids of the peroxisomal membrane are synthesized on the ER. Four potential means of accomplishing an incorporation of lipids into the peroxisomal membrane may be recognized.

1. Synthesis of the lipids directly in the membrane of the developing peroxisomes (budding ER vesicles). This means presents the problem of removal of the phospholipid synthesizing enzymes from the maturing peroxisomal membrane.

*As of late-1982.

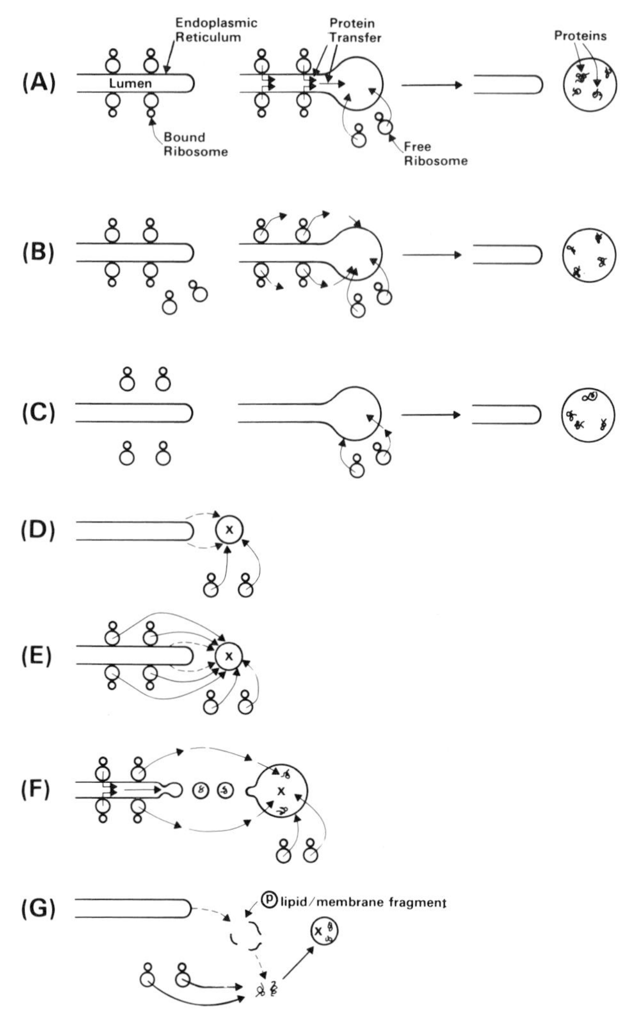

Fig. 5.13. Some possible models for organization of peroxisome biosynthesis and transfer of the component parts to the developing peroxisome. Solid line arrows indicate protein movement; dashed line arrows represent lipid movement. In the absence of arrows, movement is through the membrane continuum. In A–C, biogenesis of the peroxisomes is diagrammed as occurring by evagination of the endoplasmic reticulum. In D–G, an organizing center (X) separate from the endoplasmic reticulum is invoked. The product shown on right can be mature peroxisome, or an immature peroxisomal vesicle which will fuse with other vesicles or acquire extra enzymes (synthesized by free polysomes) and membrane lipid and protein components. Other models are possible based on currently available data.

2. Diffusion of the lipids along the ER to a budding vesicle (to become a peroxisome directly or a transfer vehicle). Diffusion of lipids is, of course, one of the basic premises of current membrane models. If the peroxisomes are synthesized as a specialized section of the ER, such diffusion into and throughout the synthesis area might be expected. It might not be so simple, however, since in castor bean endosperm the leaflet distribution of phospholipids in the glyoxysomal membrane is different from that of the ER (Cheesbrough and Moore, 1980).

3. Transfer of the lipids from the ER to the peroxisomes by transfer proteins. Evidence for the presence of phospholipid transfer proteins in the castor bean endosperm does exist (Yamada *et al.*, 1978). Such transfer proteins could be involved in conducting the phospholipids across the cytosol from the site of synthesis (the ER) to the developing glyoxysomal membranes. However, the *in vivo* transfer activity cannot be estimated currently.

4. Random collisions between the ER and peroxisome followed by subsequent capture of new portions of membrane and/or proteins by the peroxisomes. Direct contact between developing peroxisomes and the ER observed in electron micrographs (Figs. 5.3 and 1.3) may relate to such transfers of material from the ER to the peroxisomes. Evaginations (Fig. 5.3) generally are interpreted as forming peroxisomes. However, the evaginations as well as the juxtaposed views (Fig. 1.3) might also be interpreted as peroxisomes having fused with the ER temporarily, thus allowing transfer of lipids (or even proteins) through a membrane continuum. It is of interest that the juxtapositions are commonly observed in relatively undifferentiated tissues (root tips, parenchyma cells, etc.) but their association has never been adequately explained. Views of ER evagination are rare among the myriad of plant and animal cells that have been examined by electron microscopy.

3. Assembly of Proteins

The peroxisomal enzymes that have been examined are synthesized in the cytosol and not on the ER. Apparently, these proteins are not synthesized and segregated into the peroxisomes by a mechanism according to the signal hypothesis. If a newly formed enzyme, such as malate dehydrogenase, is modified at all, the modification is carried out posttranslationally rather than cotranslationally. Posttranslational transport of protein subunits into mitochondria and chloroplasts which process larger precursors into the native protein subunits is well documented (Maccecchini *et al.*, 1979; Chua and Schmidt, 1978).

Newly formed enzymes could penetrate the membrane of a peroxisome (or budding ER vesicle) without covalent modification by a mechanism described as the "trigger hypothesis" (Wickner, 1979). In this hypothesis, the transmembrane passage requires no specific signal sequence; the protein simply assumes a certain configuration capable of passing through the membrane upon being "triggered" to do so by having a "trigger" polypeptide sequence that can recognize the proper membrane. Once through the membrane, or in the proper position within the membrane, the protein configuration changes (e.g., monomer to become polymer or a change in the immediate environment) so that the "trigger" sequence ceases to exist. If this mechanism is involved in the transmembrane entry of proteins into a peroxisome or ER vesicles, some immediate questions follow. They include whether (1) the enzyme subunit alone or the whole enzyme molecule, (2) the polypeptides alone or together with the cofactor (e.g. catalase polypeptide and the heme), and (3) the protein alone or with phospholipid molecule (as has been proposed for malate synthase by Kindl, 1982) gets into or through the membrane. The capacity of the glyoxysomes and cytosolic factors in processing the newly formed proteins such as malate dehydrogenase awaits an answer.

The discussions so far in this monograph and in published reports have centered only on the entry of matrix or peripheral membrane proteins into or across the membrane of the peroxisomes or ER membrane. Another option, requiring no transmembrane movement, is that the proteins aggregate first before a new membrane encircles them, or that the proteins aggregate among themselves and with phospholipids or small membrane fragments at the same time. This mechanism requires no membrane entry and could produce either a normal size peroxisome or small vesicles which would fuse together to form a peroxisome.

IV. BIOSYNTHESIS OF PEROXISOMES IN LEAVES

The biosynthesis of leaf peroxisomes has been studied much less extensively than that of glyoxysomes, and most studies have centered on the formation of leaf-type peroxisomes in relation to the disappearance of glyoxysomes in greening cotyledons of oilseeds (Section V).

In the primary leaves of bean seedlings before greening, the cells contain small peroxisomes (0.3 μm in diameter) having a close association with the ER (Fig. 2.12). After the leaves have turned green, the peroxisomes become larger (1.5 μm) and are located near the chloroplasts (Gruber *et al.*, 1973). The enlargement is associated with severalfold increases in the activities of peroxisomal enzymes. It was suggested that the preexisting peroxisomes enlarge by fusion with ER vesicles which presumably contain the appropriate enzymes and membrane components. This suggestion could explain the findings that in the primary leaves of wheat seedlings, the individual enzymes of leaf peroxisomes are independently regulated and may have different patterns of development (Feierabend and Beevers, 1972a,b). Furthermore, the leaf peroxisomes in mature primary wheat leaves have an equilibrium density and specific enzyme activities higher than those of the peroxisomes in younger primary leaves. With *Lens* leaves, the poly(A)-mRNA directed *in vitro* synthesis of glycolate oxidase produces enzyme subunits of molecular weight 43,000, exactly the same as those isolated from the tissue (Kindl, 1982). These findings and suggestions are in accord with the latest findings on the dynamic nature of peroxisome biosynthesis (Section III).

V. BIOGENETIC RELATIONSHIP BETWEEN GLYOXYSOMES AND LEAF-TYPE PEROXISOMES IN GREENING COTYLEDONS OF OILSEEDS

In lipid-storing cotyledons of seedlings nearing depletion of the stored lipid, a change from heterotrophic to autotrophic growth occurs in response to light. This situation occurs in species of many families, and the species that have been studied most extensively are sunflower, mustard, watermelon, pumpkin, and cucumber. During the transition period, the peroxisome population changes from glyoxysomes to leaf-type peroxisomes. Figure 5.14 shows a typical response in the cotyledons of cucumber seedlings grown under a 12/12 hour light–dark cycle (Trelease *et al.*, 1971). The activities of glyoxysomal enzymes peak at about day 3 then drop to about 10% of the peak activity by day 6. Meanwhile, glycolate oxidase, a major enzyme of leaf-type peroxisomes (but also found in glyoxysomes at a low level), does not increase rapidly until after day 3, and approaches a plateau from day 5 onward. Catalase, a major enzyme of both types of peroxisomes, peaks with the glyoxysomal enzymes, then decreases, but does not continue to follow the decline pattern of other glyoxysomal enzymes. Rather, it levels off as does the increase in glycolate oxidase after day 6. The transition obviously is a complex system, and at any given time some representatives of both peroxisome types occur together in the same tissue.

One of the more intriguing questions concerning peroxisomal ontogeny is whether glyoxysomes can be converted to leaf-type peroxisomes in these cotyledons. A positive answer to this question represents the "interconversion model." On the contrary, the so-called "two-popu-

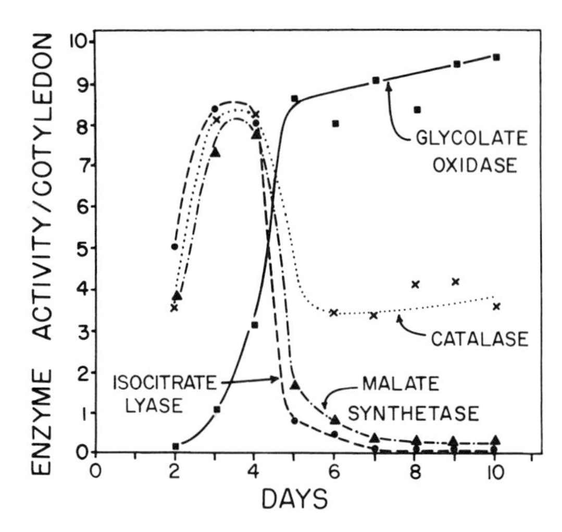

Fig. 5.14. Changes in glyoxysomal (isocitrate lyase, malate synthase, catalase) and leaf-type peroxisomal (glycolate oxidase, catalase) enzymes in cucumber cotyledons of seedling grown under a light–dark (12/12 hr) cycle. From Trelease *et al.* (1971).

lation model" indicates that an entirely new population of leaf-type peroxisomes is synthesized and the old glyoxysomes are destroyed, with no part being directly reused. In experimental approaches, the relative proportion of the two types of peroxisomes can be controlled to some extent by manipulating the light regime, but overlap between the two types still exists which makes it difficult to interpret experimental results. Furthermore, the two peroxisome types are quite similar in size and buoyant density, and it has not been possible to adequately separate them from each other following sucrose gradient centrifugation of tissue extracts. The maximum reported equilibrium density difference between the two peroxisome types from watermelon cotyledon is 0.007 ± 0.004 out of 1.25 g/cm^3 (Gerhardt, 1973). Thus, the work in this area involves interpreting experimental data utilizing a mixture of the two peroxisome types.

A. Interconversion Model

In 1971, Trelease *et al.* first suggested the interconversion model based on electron microscopic evidence. In greening cucumber cotyledons, the estimated numbers of peroxisomes (microbodies) per cell do not decrease in proportion to the decline in glyoxysomal enzyme activities, and no evidence for new synthesis or destruction of whole peroxisomes can be observed. It was suggested that cytoplasmic invaginations in the peroxisomes (Fig. 2.16) observed only during the changeover period might represent a process associated with the gain of leaf-type peroxisomal enzymes and/or loss of glyoxysomal enzymes.

Another line of evidence to support the interconversion model comes from experiments utilizing heavy isotope density labeling. The density label, deuterium, was administered as ^2H_2O to the tissue during the period of leaf-type peroxisome appearance in order to increase the density of this organelle. The deuterium is first incorporated into the amino acids, lipids, and carbohydrates by the cells, which then use these materials to synthesize proteins, etc., to be used in forming new organelles. Existing organelles should not incorporate these metabolites and, thus, the deuterium to a large extent. Thereby, newly synthesized organelles should be more dense than preexisting organelles, and those composed of some new and some old materials would be intermediate in density. When this technique was applied to cotyledons of both sunflower and cucumber during a period of light stimulation of leaf-type peroxisome appearance, very small increases (0.01–0.02 g/cm^3) in density of the organelles from the water controls was observed. These data were interpreted (Theimer *et al.*, 1975) to mean that the bulk of the leaf-type peroxisomes are derived from preexisting materials. A similar density labeling technique has been applied specifically to catalase in sunflower cotyledons (Betsche and Gerhardt, 1978), in order to determine whether it is synthesized anew during the appearance of leaf-type peroxisomes. Through the use of ^2H_2O-labeling combined with CsCl density gradient centrifugation to estimate the densities of the enzyme (Fig. 5.15), the amount of catalase synthesized during peroxisome appearance may be estimated. Little or no shift of the catalase density occurs in the sunflower cotyledons during the period when peroxisome enzyme activity increase rapidly. A control chloroplast enzyme, NADP-linked glyceraldehyde-3-phosphate dehydrogenase, increases in density about 1%, demonstrating that the density label is incorporated into at least some of the proteins of the tissue. From these data it may be argued that the catalase is held over from glyoxysomes to leaf-type peroxisomes. However, these data and their interpretation could have allowed for considerable error (up to 20% of the catalase could have been newly synthesized), and so the conclusion is not unequivocal. Furthermore, and of even more importance, is the finding that the syntheses of glycolate oxidase and catalase are not closely coupled during leaf-type peroxisome biogenesis (Feierabend and Beevers, 1972a). In leaves of wheat seedlings, glycolate oxidase increases rapidly as a response to light, while catalase activity has a delayed increase (1–2 days lag) as a response to the same treatment (Fig. 5.16). Thus, it is possible that leaf-type peroxisomes with little or no catalase are synthesized as a response to light.

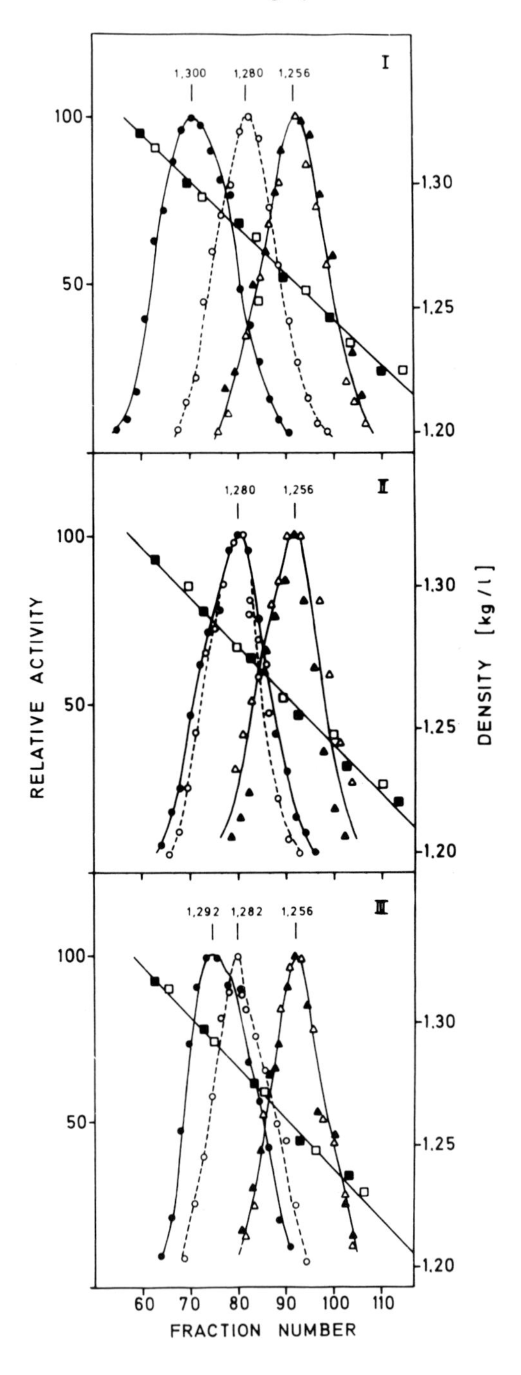
I
1,300
1,280
1,256
II
1,280
1,256
III
1,292
1,282
1,256
100
50
1,30
1,25
1,20
RELATIVE ACTIVITY
DENSITY [kg / l]
60
70
80
90
100
110
FRACTION NUMBER

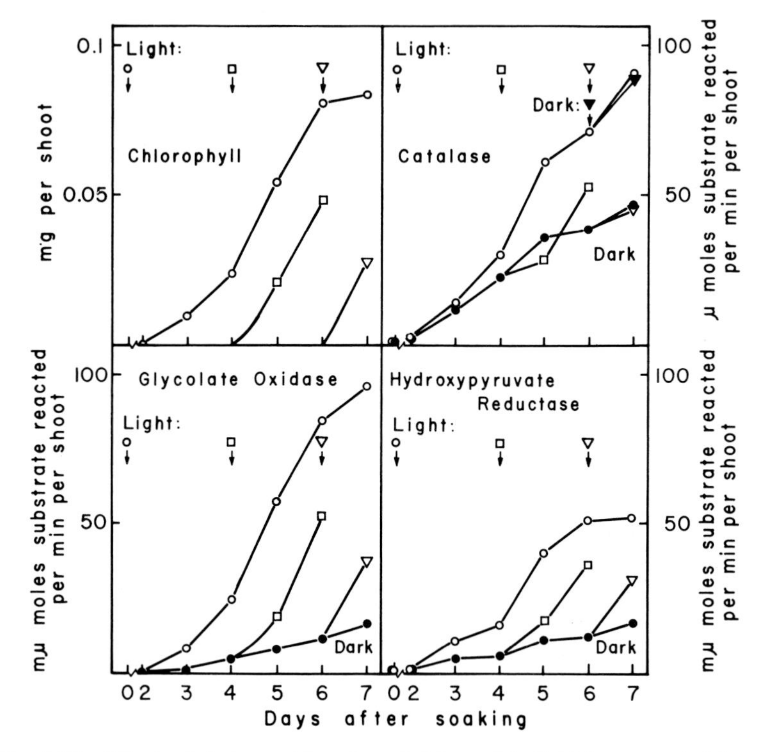

Fig. 5.16. Effects of darkness (closed symbols) or white light (open symbols) on chlorophyll content and selected peroxisomal enzymes in wheat leaves. Plants were grown in continuous darkness (●), continuous light (○), or placed from darkness into light after 4 days (□) or 6 days (▽). In some cases plants grown in continuous light were placed into the dark after 6 days (▼). From Feierabend and Beevers (1972a).

Further evidence in favor of the interconversion hypothesis comes from estimates of the percentage of microbodies cytochemically stained for malate synthase (glyoxysomal) or glycolate oxidase (leaf-type peroxisomal) activity (Burke and Trelease, 1975). When cucumber cotyledons are sampled under germination conditions in the presence of light such that intermediate levels of both enzymes are present, virtually all the microbodies observed *in situ* stain for both enzymes. Following isolation, 80–90% of the identifiable microbodies can be stained for both

---

Fig. 5.15. Distributions of catalase from unlabeled (○) and density labeled (●) sunflower cotyledons on isopycnic CsCl gradients. The standard was lactate dehydrogenase (△,▲). Separation I was from sunflower labeled during active glyoxysome synthesis, II from a dark to light transition period, and III from sometime after the dark to light transition period. From Betsche and Gerhardt (1978).

activities (Figs. 5.17 and 5.18). These data provide evidence for an enzyme changeover occurring within an ongoing population of peroxisomes, since both enzymes appear to have occurred in the same organelle. However, the clarity of the pictures (Figs. 5.17 and 5.18), the identity of the microbodies, and the susceptibility of the method to subjectivity and difficulty in sample collection have been criticized (Beevers, 1979). Furthermore, the *in situ* cytochemical observation would have been more convincing if the staining of the two enzymes had been performed separately on serial sections rather than on two different pieces of tissue, an approach that could not be carried out on the cytochemistry of the two enzymes.

B. Two-Population Model

The major evidence in favor of the two-population model derives from work utilizing watermelon seedlings (Kagawa *et al.*, 1973a,b; Kagawa and Beevers, 1975). When the seedlings are maintained in continuous darkness, a major loss of protein from the glyoxysomal fractions isolated from the cotyledons occurs. Treatment with light accelerates this loss, whether it is initiated early (4 days) or late (10 days) during the postgermination period (Figs. 5.19 and 5.20). During the light treatment, the leaf-type peroxisomal enzyme activities increase dramatically. The final level of protein in the isolated leaf-type peroxisome fraction is only one-fourth of that present in the glyoxysomal fraction isolated at the peak of activity. The total catalase activity common to both types of peroxisomes behaves similarly to the total peroxisomal protein in the 4-day light-treated tissue. Assuming that each glyoxysome or leaf-type peroxisome contains the same amount of protein, these data appear to counter one of the major arguments originally used in favor of the interconversion model, i.e., that the number of peroxisomes does not decrease during the changeover from glyoxysomes to leaf-type peroxisomes (Trelease *et al.*, 1971). However, since the protein in the peroxisomal fractions does not entirely disappear even during prolonged darkness, it still might be argued that some portion of the glyoxysomes remained to be transformed. Also, some protein degradation may be expected in the glyoxy-

---

Fig. 5.17. Representative peroxisomal fraction isolated from cucumber cotyledons incubated in a complete reaction mixture for malate synthase reactivity (top) or a control medium (minus substrate) (bottom). The cotyledons were obtained from seedlings grown during the dark-to-light transition. Approximately 83% of the particles considered to be microbodies (peroxisomes) were stained. p, plastid; pb, protein body fragments. Bars = 1 μm. From Burke and Trelease (1975).

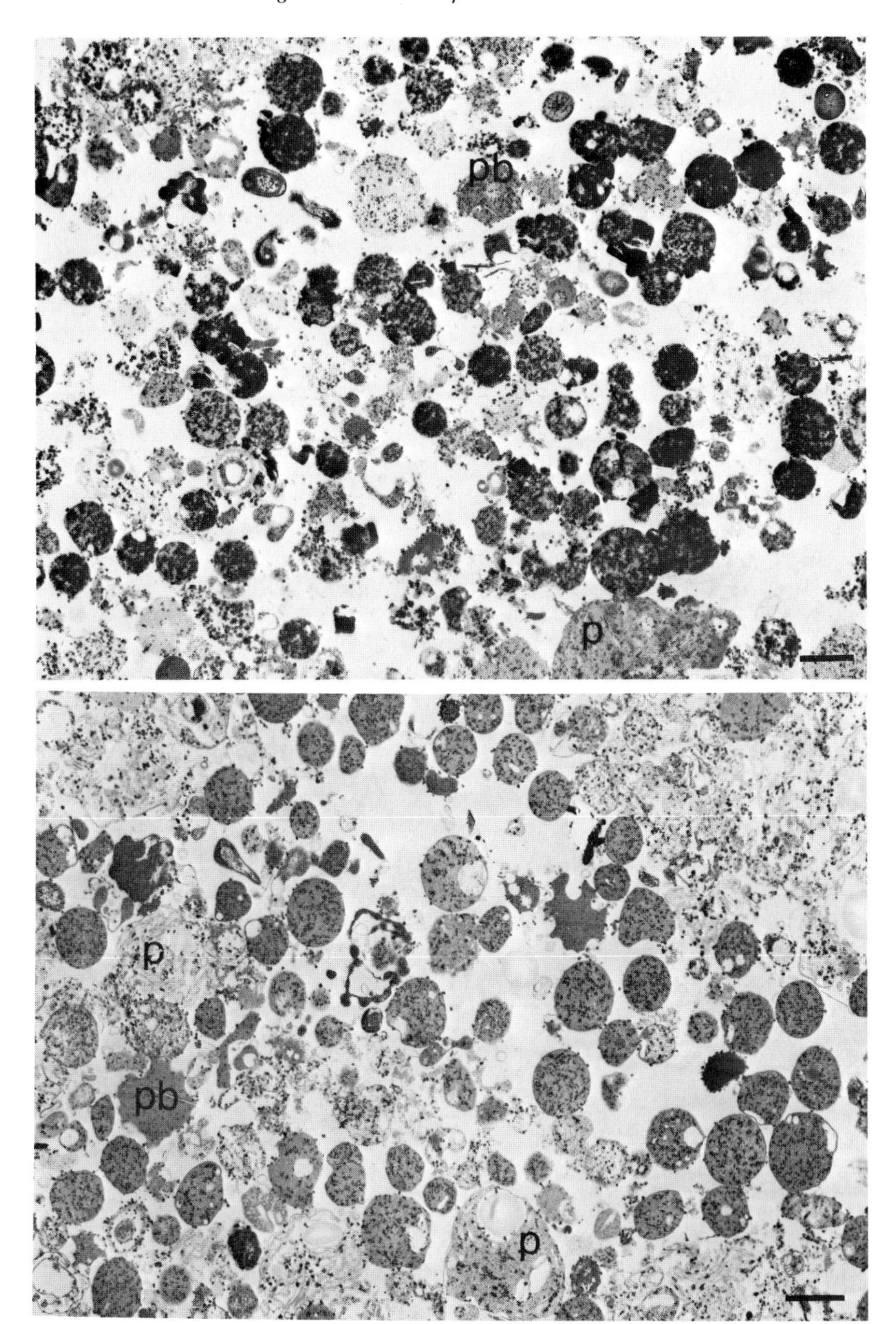
pb
p
p
pb
p

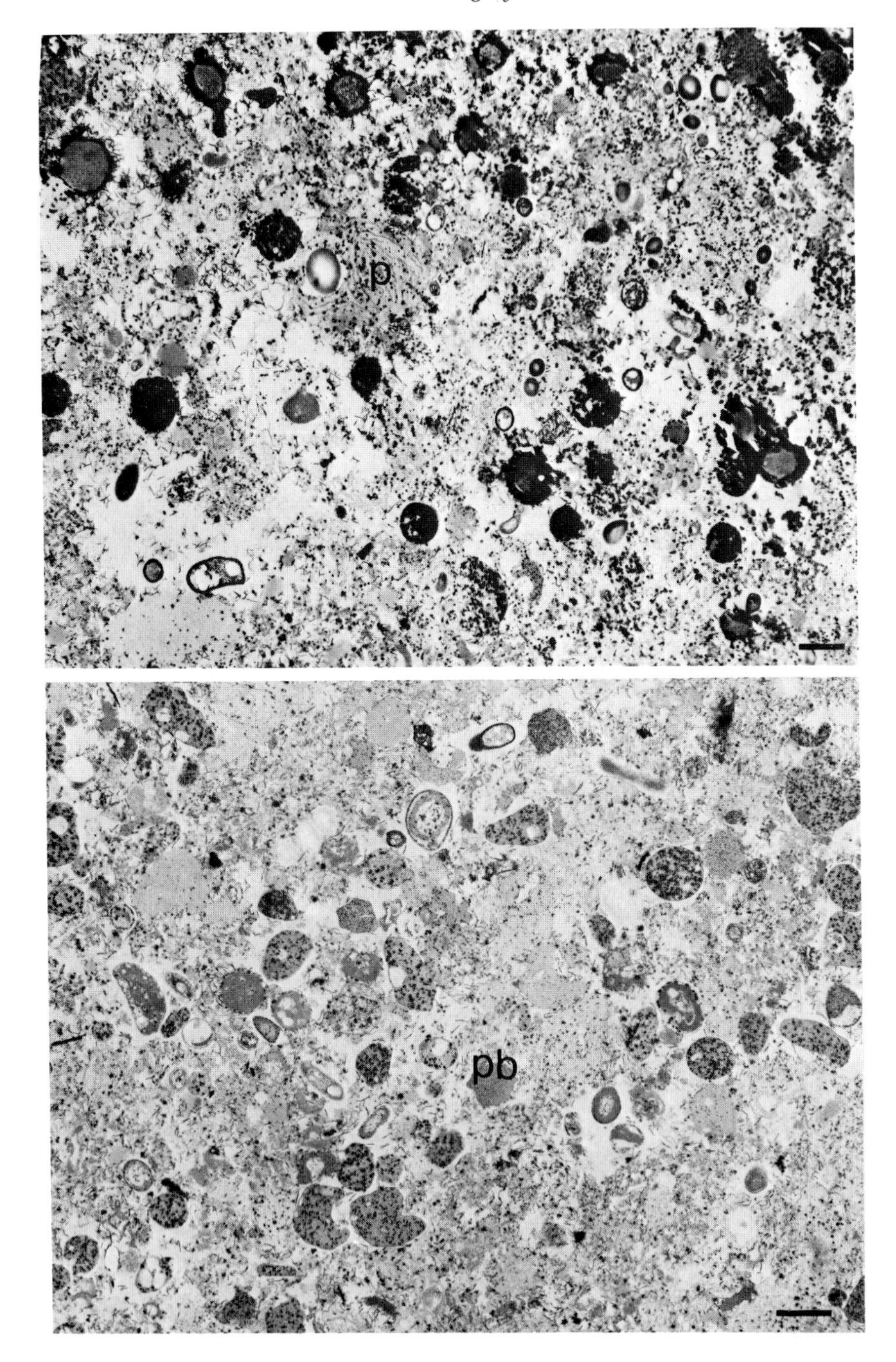
p
pb

somes in the interconversion model, since glyoxysomal enzymes are envisioned as being removed during conversion. One additional criticism is that protein body fragments might contaminate the glyoxysomal fraction, as is true for many other tissues (Becker *et al.*, 1978; also Chapter 3), and that the protein measured is due substantially to this contribution. If so, the increased loss of storage proteins might produce an artificial decrease in "glyoxysomal" proteins.

The watermelon system also has been examined for possible membrane lipid changes on the assumption that the phospholipids of the peroxisomal membranes would be expected to remain the same during the transformation of glyoxysomes to leaf-type peroxisomes (Kagawa *et al.*, 1975). If such were the case, then the phospholipids of the peroxisomal fraction labeled during glyoxysome synthesis with radioactive choline, containing $^{14}C/^{3}H$ at a known ratio, should have that ratio unchanged even though ^{14}C-choline is fed to the tissue during the transformation period. Contrary to that expectation, however, there was a distinct increase in the $^{14}C/^{3}H$ ratio in membrane lipids of the peroxisomal region of the sucrose gradient used to isolate organelles from watermelon extract (Fig. 5.21). This suggested new membrane synthesis in that fraction. The major difficulty in interpreting these data lies in the fact that changes in contaminating membrane lipids in the peroxisomal fraction could contribute extensively to the results. Intact plastids are a definite candidate for contamination in that region of the gradient and the data presented did not adequately rule out their presence. Another potential explanation of the data against having two populations is that the rise in the $^{14}C/^{3}H$ ratio may have been the result of an accelerated turnover (degradation with subsequent replacement) of peroxisomal membranes as a response to the light treatment. Modification of the membranes during the theorized transformation period may also have occurred.

Evidence in favor of the two-population model also includes density labeling experiments utilizing deuterium. In experiments similar to those of Theimer *et al.* (1975) described above, but using pumpkin, Brown and Merrett (1977) observed also a small density shift (0.01 g/cm^3, within the range observed by Theimer *et al.*) during the ap-

---

Fig. 5.18. Representative peroxisomal fraction isolated from cucumber cotyledons incubated in a complete reaction mixture for glycolate oxidase reactivity (top) or a control medium (minus substrate) (Bottom). The cotyledons were obtained from seedlings grown during the dark-to-light transition. Approximately 87% of the particles considered to be microbodies (peroxisomes) were scored as stained. p, plastid; pb, protein body fragments. Bars = 1 μm. From Burke and Trelease (1975).

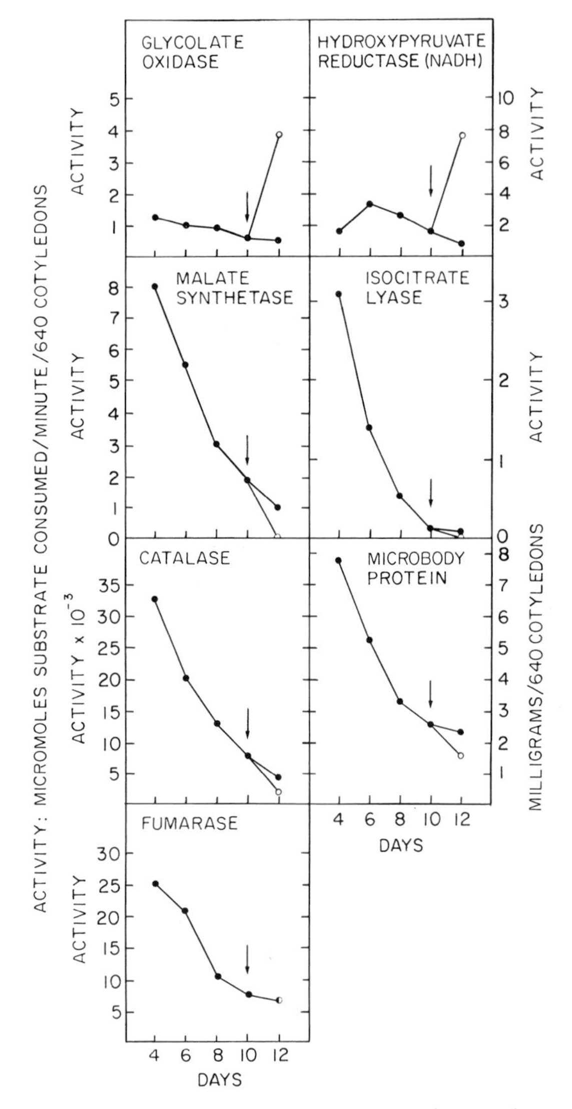

Fig. 5.19. Changes in the activities of glyoxysomal (malate synthase, isocitrate lyase, catalase), leaf-type peroxisomal (glycolate oxidase, hydroxypyruvate reductase, catalase), and mitochondrial (fumarase) enzymes and peroxisomal protein in continuous dark (○) or following treatment with light after 10–12 days of darkness (●). Arrow indicates the age at which the seedlings were transferred to light. From Kagawa and Beevers (1975).

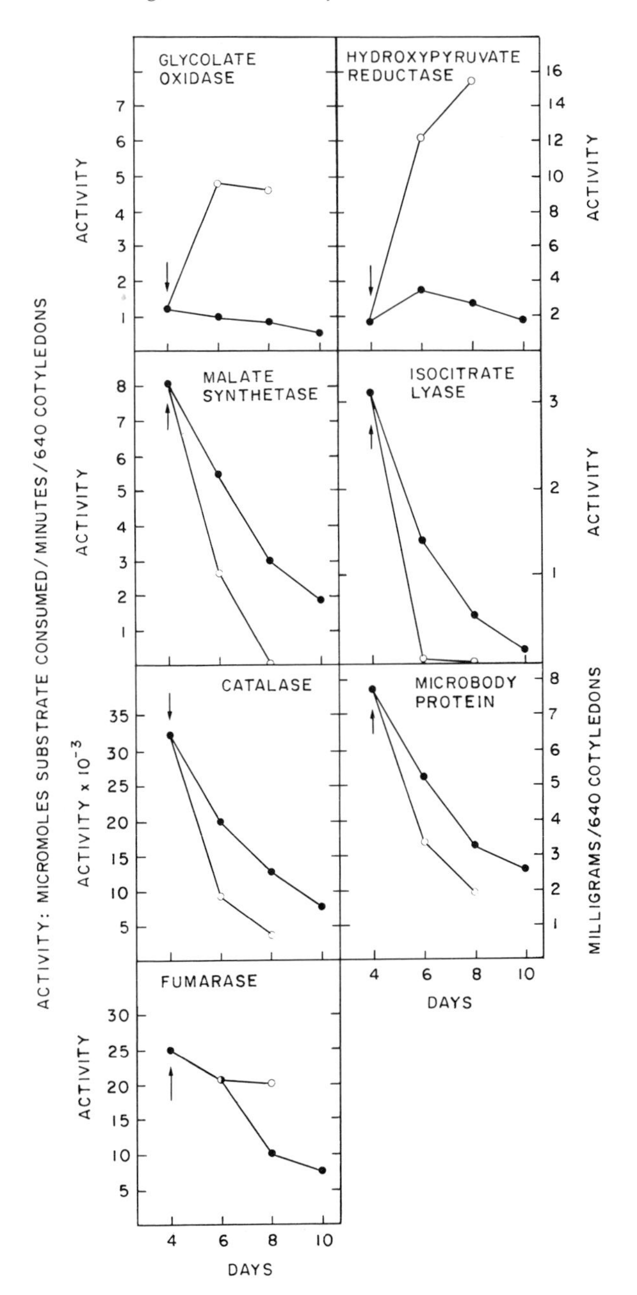

Fig. 5.20. As in Fig. 5.16, except light treatment was at about day 4 and development was allowed to proceed only until day 10. From Kagawa and Beevers (1975).

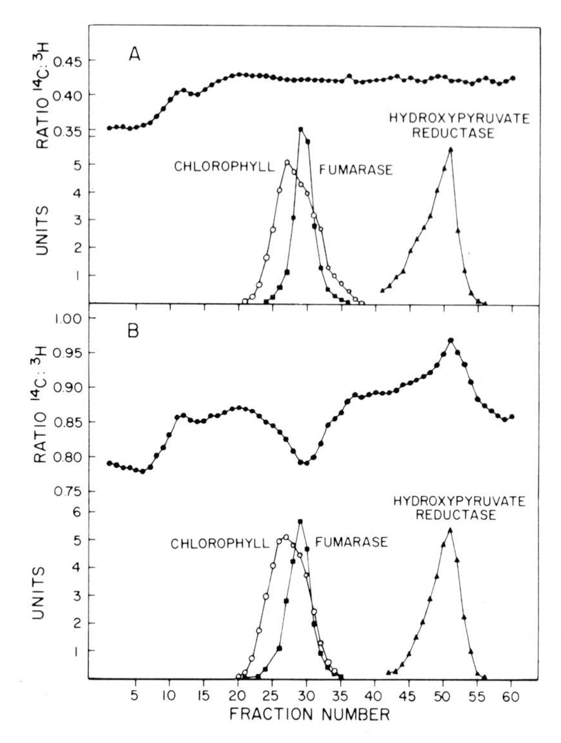

Fig. 5.21. The ratio of $^{14}C:^{3}H$ (●) in organelle fractions from cotyledons of watermelon seedlings separated on a sucrose density gradient. (A) The cotyledons were given [^{14}C-^{3}H]choline in the dark at 0 hr, transferred to light at 12 hr, and [^{12}C]choline was added at 18 hr. (B) As (A) except that [^{14}C]choline was reintroduced at hour 18. The tissue was harvested after 30 hr in both cases. Chlorophyll, fumarase, and hydroxypyruvate reductase were used as markers for chloroplasts, mitochondria, and peroxisomes, respectively. From Kagawa *et al.* (1975).

pearance of peroxisomes as a response to light. Contrary to the conclusion of Theimer *et al.* (1975), they calculated that this shift was on the order expected if all the proteins of the peroxisomes were newly synthesized rather than a modification of the glyoxysomes into leaf-type peroxisomes.

C. Transient-Intermediate Model

A compromise model ("transient-intermediate" model) has been set forth which proposes the existence of a transient form of peroxisome intermediate between the glyoxysomes and leaf-type peroxisomes (Fig. 5.22). The transient form is proposed to occur during the changeover between the two major peroxisome populations and represents a form in which both the remnants of glyoxysomal enzyme synthesis and the beginnings of leaf-type-peroxisomal enzyme synthesis occur; enzymes of each type would be incorporated into the same organelle (Schopfer *et al.*, 1976a). This model is based on ultrastructural observations of cotyledons of mustard seedlings grown under varied white and far-red light irradiations. During the transient period when enzyme activities of both types of peroxisomes are present, the "intermediate" peroxisomes were found associated with both plastids and lipid bodies. In leaves, peroxisomes commonly are seen associated with plastids, whereas in lipid-storing endosperm and cotyledons their association with lipid bodies is not unusual. Supporting evidence comes from studies with cucumber cotyledons (Koller and Kindl, 1978). Upon illumination of 3-day-old seedlings, both glycolate oxidase and malate synthase increase

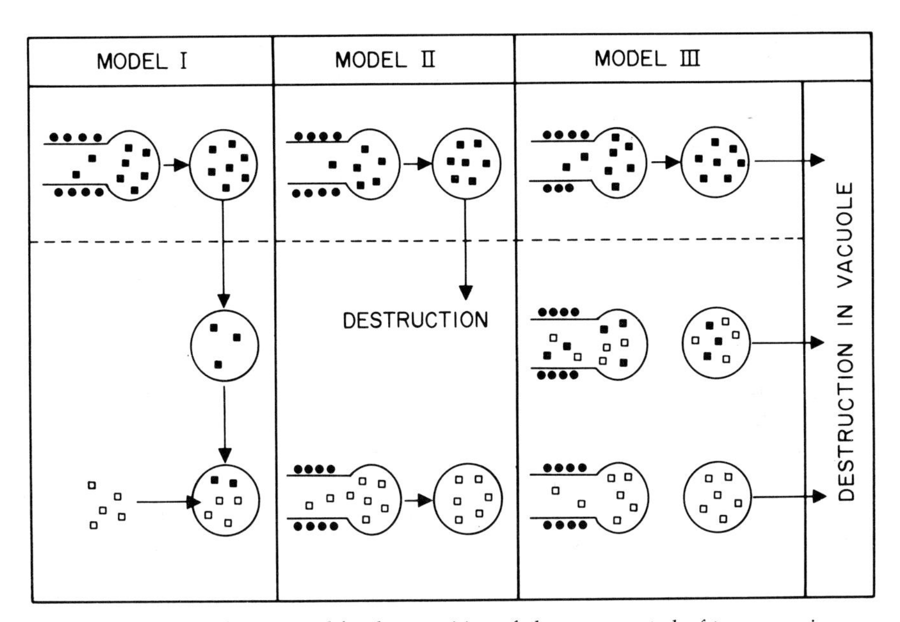

Fig. 5.22. Models proposed for the transition of glyoxysomes to leaf-type peroxisomes in greening cotyledons of oil seeds. Phase I, above the dashed line indicating the onset of light, shows glyoxysome formation. Vesiculation of a portion of the endoplasmic reticulum is shown. ●, ribosomes; ■, glyoxysomal enzymes. Model I, the interconversion model, shows the loss of some glyoxysomal enzymes and the insertion of leaf-type peroxisomal enzymes (□). Model II, the two-population model, shows the destruction of glyoxysomes and the *de novo* biogenesis of leaf-type peroxisomes initiated by light. Model III is the compromise model. This figure illustrates the essential ideas of the models, even though the synthesis of glyoxysomal or leaf-type peroxisomal enzymes solely from the rough endoplasmic reticulum is not valid any more. From Beevers (1979).

in the isolated peroxisomal fractions. The increase in malate synthase is due to *de novo* synthesis, and persists for 15 hr after illumination when already there has been a drastic increase in glycolate oxidase.

It should be noted that the compromise model requires a massive destruction of the glyoxysomes as does the two-population model. Indeed, the intermediate peroxisomes would represent only transient organelles expected during the changeover, in either the two-population model where both syntheses occur simultaneously, or in the one-population model if the organelles do not turnover rapidly. One concern for the compromise model is the same as for the two-population model, in that glyoxysome destruction should be apparent, but is not observed with the electron microscope. It also is likely that there are changes in the plastids and cytoplasm during this period and such changes might affect the affinity of the peroxisomes for other organelles. For example, the phosphate levels are known to affect the association between peroxisomes and plastids *in vitro* (Schnarrenberger and Burkhard, 1977).

D. Assessment

In assessing the available information, no equivocal evidence has been presented to support any one of these models. The number of peroxisomes (microbodies) per cell as observed with the electron microscope before and after the greening process is a very basic piece of information that is lacking. The two-population model, as well as the compromise model, has been based heavily, but not necessarily exclusively, on the ER-vesiculation model of peroxisome biogenesis from budding of ER. This ER-vesiculation model of peroxisome biogenesis is rapidly losing support in view of the latest findings (Section III) that the assembly of peroxisomal enzyme components is much more dynamic. Furthermore, the differential development of peroxisomal enzymes during seed maturation and germination, as demonstrated in cotton and cucumber cotyledons (Section III,B), indicates that addition of enzymes to preexisting peroxisomes is possible. Similar supportive evidence also comes from the biogenesis of leaf peroxisomes in wheat and bean leaves (Section IV). These recent discoveries provide possible examples of enzyme addition to preexisting peroxisomes and emphasize the dynamic nature of peroxisome biogenesis. Although these findings are in agreement with the interconversion model, they are not evidential. With the current techniques developed for the study of peroxisome biogenesis (Section III,A,2), the problem can be tackled more directly. For example, a possible demonstration of glycolate oxidase (and hydroxypyruvate reductase) being synthesized on free polysomes and incorporated into isolated glyoxysomes/peroxisomes would favor the interconversion model. More studies are warranted.

VI. BIOGENESIS OF PEROXISOMES IN MICROORGANISMS (YEASTS)

When many yeast species, such as those of *Hansenula, Candida,* and *Kloeckera,* are cultured with certain carbon sources such as methanol, acetate, ethanol, and fatty acids, etc., numerous peroxisomes with specialized metabolic functions are induced (Chapter 4). There have been several studies on the biogenesis of peroxisomes in these yeast cells after their transfer from glucose-to methanol-containing medium. Based mostly on electron microscopic observations, two contrasting hypotheses have been proposed. The "*de novo* hypothesis" states that the peroxisomes are produced by the formation of dense cores in the cytosol followed by encasement with a limiting membrane derived from the ER (Tsubouchi *et al.,* 1976). In contrast, the growth-and-division hypothesis states that larger peroxisomes containing methanol oxidase are derived from the original peroxisomes (presumably the "unspecialized" peroxisomes) by the deposition of methanol oxidase to form the crystalline inclusions (Tanaka *et al.,* 1976; Veenhuis *et al.,* 1979). In budding cells, the mature, enlarged peroxisomes in the mother cells divide and the small peroxisomes are transferred to the developing buds (Fig. 5.23). In the new cell, the small peroxisomes increase in size but there is no increase in the number per cell. If the cells are allowed to age, the enlarged peroxisomes will divide to produce more peroxisomes of small sizes (Fig. 4.17). In these studies, no association of peroxisomes with ER has been reported.

VII. DEGRADATION

Most studies of peroxisome ontogeny have been devoted to peroxisome biogenesis rather than degradation, and little direct information is available on this latter aspect. In castor bean seedlings, the endosperm becomes senescent after the food reserve has been depleted. Under this condition, microbody-like structures inside a single-membraned, potentially digestive, sac have been observed *in situ*(Fig. 5.24;Vigil, 1970). Such electron micrographs are rare in the literature, and confirmation of the digestive nature of the sac (e.g., by cytochemistry of hydrolytic enzymes) is lacking. Leaf peroxisomes in bean primary leaves have been

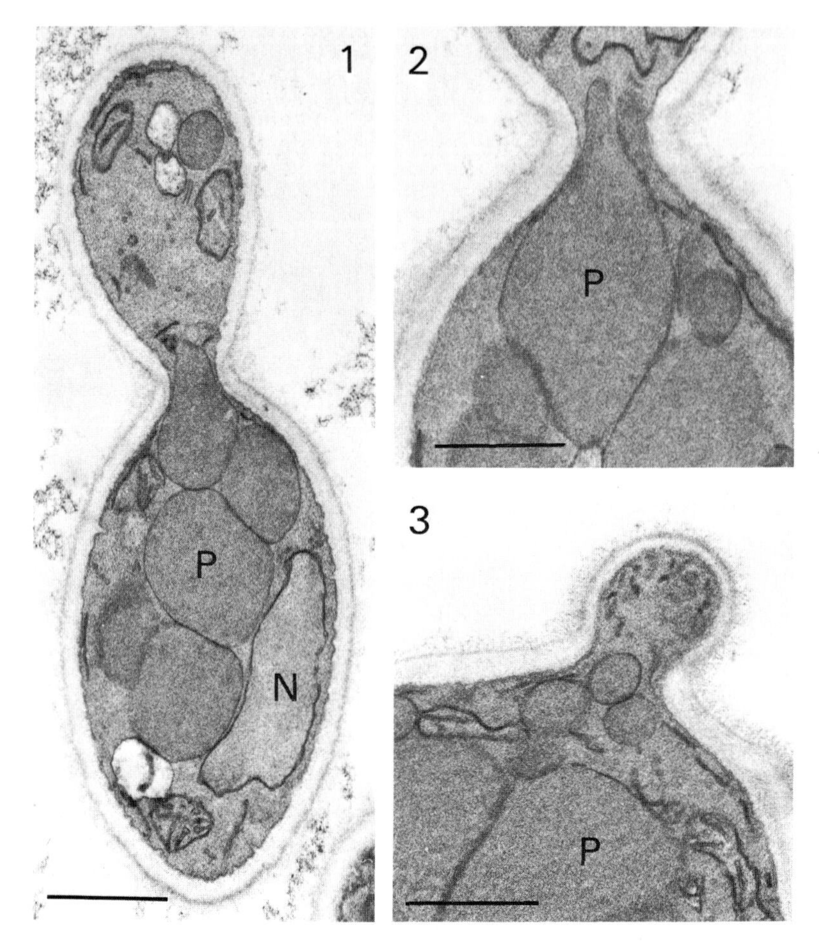

Fig. 5.23. Peroxisomes in *Hansenula polymorpha* grown on methanol. (1) A dividing cell with a peroxisome in the neck. (2) A peroxisome in the neck between mother cell and bud. (3) Migration of a small peroxisome into the developing bud. p, peroxisome; n, nucleus. Bars = 0.5 μm. From Veenhuis *et al.* (1978).

proposed to be long-lived on the basis of the absence of any obvious dissolution, even in older leaves (Gruber *et al.*, 1973). In yeasts, the peroxisomes may be long-lived also (Section VI). In some cases, the peroxisomal enzyme activities disappear at similar rates, but detailed investigations of this coordination of degradation have not been performed. Little is known about peroxisome degradation in nonplant tissues, either.

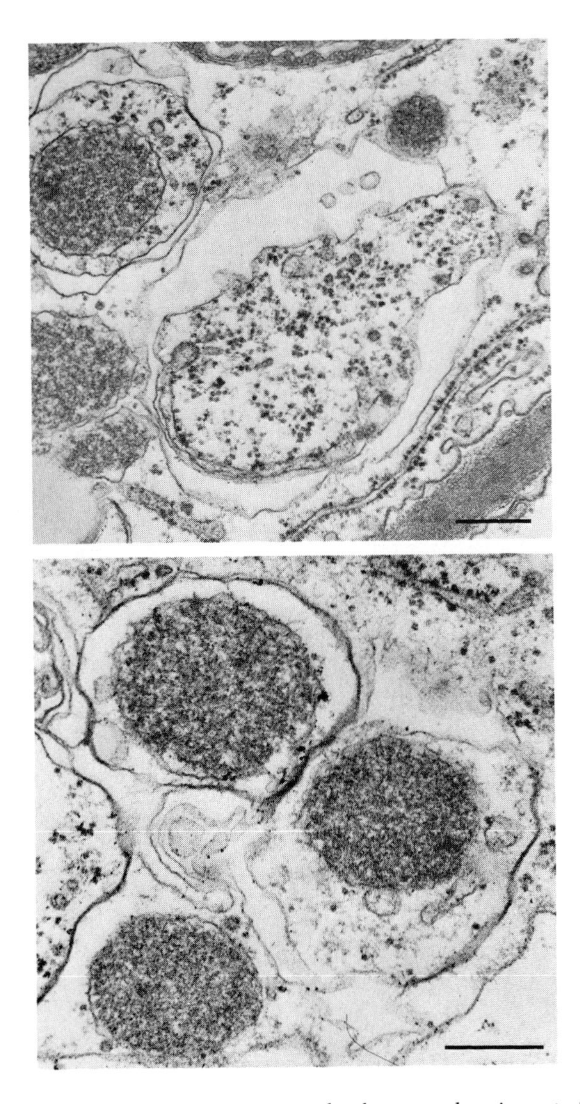

Fig. 5.24. Segregation of glyoxysomes and other cytoplasmic material by double-membraned cisternae in endosperm cells of castor bean seedlings. (Top) The limiting membranes of the larger vacuole with a segment of rough ER and many free ribosomes appear separate, while those of the smaller vacuole containing a microbody (glyoxysome) and a few ribosomes are partially fused. (Vigil, 1970). Bar = 0.5 μm. From Vigil (1970). (Bottom) Microbody (glyoxysome) in a vacuole (top), glyoxysome and other cytoplasmic material in a vacuole (middle), and glyoxysome in cytoplasm (bottom). Bar = 0.5 μm. From E. Vigil (unpublished).

VIII. SUMMARY

In the ER-vesiculation model, glyoxysomes were conceived to be synthesized by the budding off of specialized vesicles of the endoplasmic reticulum to become a mature peroxisome; the glyoxysomal enzymes would be synthesized on bound polysomes and cotranslationally segregated into the ER lumen. Recent findings are inconsistent with this model. As of what we know by late-1982, the phospholipids of the glyoxysomal membrane are synthesized on the endoplasmic reticulum whereas the membrane peripheral proteins and matrix proteins are synthesized in the cytosol. Only malate dehydrogenase is known to contain an extra polypeptide sequence in the newly synthesized enzyme, and this sequence is cleaved posttranslationally, not cotranslationally. The biosynthetic site of membrane integral proteins is unknown. The newly synthesized phospholipids and proteins are assembled at an unknown site to produce either a mature peroxisome or a small membrane vesicle as an intermediate of an enlarging peroxisome. Although the biogenesis of glyoxysomes has been pursued actively, little is known about the biogenesis of leaf peroxisomes. In greening cotyledons of many oilseeds, whether the glyoxysomes are converted to leaf-type peroxisomes or there are two independent populations of organelles having separate biogenesis has not been unequivocally resolved. In yeasts, there are numerous reports based on electron microscopic observations that peroxisomes are formed by organelle enlargement and division. The mechanism of peroxisome degradation is unknown.

6

Control of Ontogeny

I. INTRODUCTION

The regulation of development of several peroxisomal enzymes has been studied. Although these investigations have not yet provided a concrete explanation of how peroxisomal enzyme activity is regulated, they point to some interesting areas of future work. Information on the control of individual enzymes can lead to understanding the control on the biogenesis of the entire peroxisomes. Because the peroxisomes have not been shown unambiguously to contain their own nucleic acid, it is reasonable to assume that the genes coding for the peroxisomal proteins reside in the nucleus. Of all the peroxisomal enzymes in higher plants, only catalase in maize is known to have its genes in the nucleus. A few peroxisomal enzymes are known to be synthesized on cytoplasmic ribosomes (80 S) and not on plastid or mitochondrial ribosomes (70 S).

II. DIFFERENTIAL EXPRESSION OF ISOZYMES

A. Catalase

The genetics of catalase in higher plants has been studied only in maize, but the available information is substantial. A complicated genetic system controls the expression and regulates the activity of this peroxisomal enzyme. Catalase isolated from diverse species of higher plants, microorganisms, and animals is a tetramer of identical subunits (Chapter 4). At least three structural genes for catalase have been found in maize (Scandalios, 1975; Scandalios *et al.*, 1980a). *Cat 1* is expressed in developing seeds and consists of six alleles. A homozygote of each allele codes for one unique isozyme (Fig. 6.1). Crosses between any two of the allelic types result in five distinct isozymes in the F_1 heterozygotes (Fig. 6.2), including the two parental types and three hybrid enzymes. Similar isozyme patterns can be obtained from a mixture of the two parental isozymes by disassociating the enzymes with high salt solution and subsequently allowing the subunits to reassociate. The findings confirm that catalase is a tetrameric enzyme. A second gene, *Cat 2,* is located on a different chromosome. It codes for a catalase isozyme that has different immunological and physical properties than the catalase isozyme of *Cat 1*. *Cat 1* is expressed during seed maturation, and *Cat 2* is expressed in postgermination. During the overlapping period when both genes are expressed, five isozymes can be observed with a changing isozyme pattern in accord with the tetrameric composition of two genetically different subunits (Fig. 6.3).

A third gene, *Cat 3,* consisting of two codominant alleles, is expressed in young leaves and pericarp of nearly developed seeds (Scandalios *et al.*, 1980a). The catalase coded by *Cat 3* is electrophoretically different from those coded by the other two genes. In a few inbred lines where *Cat 3* is expressed simultaneously with either *Cat 1* or *Cat 2* in the same tissue, the different subunits do not interact *in vivo* to form hybrid molecules.

A regulator gene, *Car 1,* located at 37 map units from *Cat 2* in the same chromosome, is expressed in the scutella (Scandalios *et al.*, 1980b). Its expression somehow reduces the synthesis but not the degradation of catalase. In an inbred line (R6-67) with a variant allele at *Car 1,* the level of catalase is elevated. Its regulatory effect is quite specific for catalase since other enzymes apparently are not affected, and the gene product is not a protease (see Fig. 6.12). The gene does not affect the development of isocitrate lyase and malate synthase activities, suggesting that it does not regulate other glyoxysomal enzymes simultaneously. In addition to

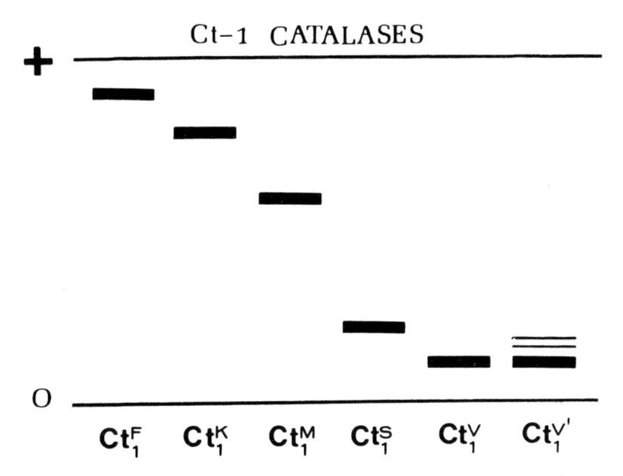

Fig. 6.1. Diagram of the results of starch gel electrophoresis (pH 7.4) of the six catalase variants found in endosperm of different inbred strains of *Zea mays* during seed maturation. The horizontal axis indicates alleles at the Ct_1 locus and 0 is the point of sample application. From Scandalios (1975).

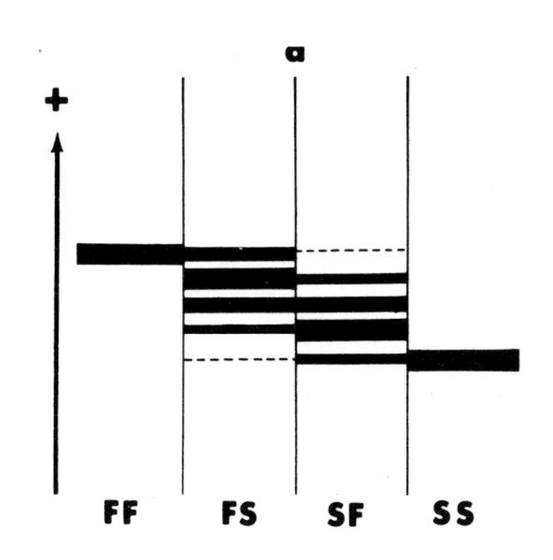

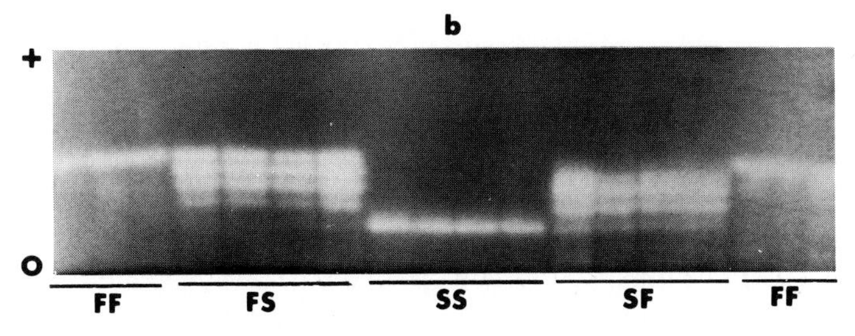

Fig. 6.2. Diagram (a) and zymogram (b) of catalase isozymes in the endosperm of two allelic variants (*F* and *S*) and the two heterozygote results from reciprocal F_1 crosses ($F \times S$ and $S \times F$) of *Zea mays*. The triploid nature of the endosperm accounts for the apparent gene dosage effects. From Scandalios (1975).

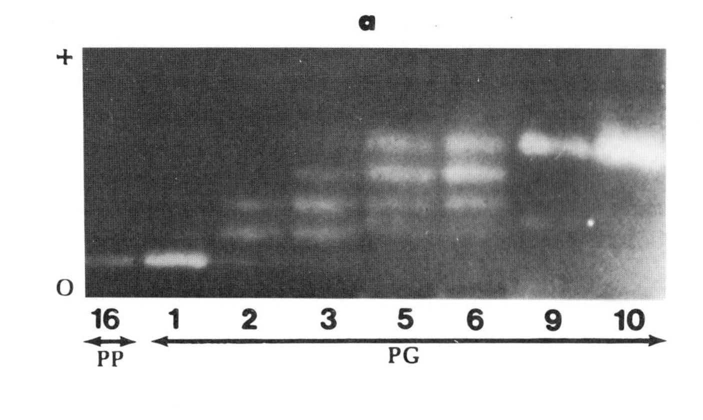

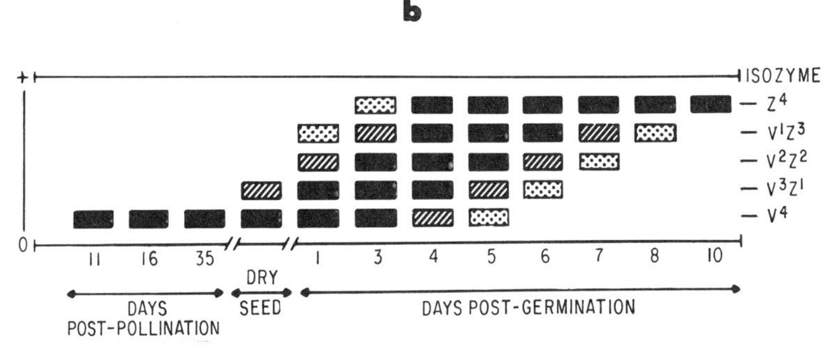

Fig. 6.3. A zymogram (a) and its diagrammatic representation (b) of the shifts in catalase isozymes from scutella extracts obtained during seed maturation and postgermination of *Zea mays*. The V^4 homotetramer is composed of subunits from the Ct_1 gene which is represented in maturing seeds and early seedlings stages. Z^4 is an expression of the Ct_2 gene and appears at the onset of germination. Intermediate isozymes are mixed subunits from both genes. PP is days postpollination, PG is days postgermination, and the numerical units are days. From Scandalios (1975).

the effect of the regulatory gene, an inhibitor of catalase activity has been characterized, but its genetics has not been studied (Sorenson and Scandalios, 1980). The inhibitor is a low molecular weight protein, and inhibits some 20% of the *in vitro* activity of catalase from maize as well as from microbial and animal sources. The subcellular location and physiological role of this inhibitor are unknown (also see Section III,D).

Catalase isozymes of both *Cat 1* and *Cat 2* were found in the glyoxysomal fraction isolated from the scutella of germinated seeds. The bulk of the enzymes was soluble, but this may reflect the difficulty in preparing organelles from the scutella. Catalase of *Cat 3* was found in peroxisomal, mitochondrial, and soluble fractions (Scandalios *et al.*, 1980a); its precise subcellular localization requires further investigation.

Temporal development of more than one catalase isozyme also occurs in the cotyledons of mustard seedlings (Schopfer *et al.*, 1976b). The catalase isozyme patterns in the cotyledons changed upon illumination with continuous far-red or fluorescent light. Only three isozyme bands are present in etiolated tissue; after exposure to light, these three isozymes remain while seven new ones appear (Fig. 6.4). All the cotyledon isozymes also are present in extracts of mature green leaves. The subcellular location of these isozymes is not known, and the effect of light on the expression of new isozymes is not understood.

B. Aspartate-α-ketoglutarate Aminotransferase

Aspartate–α-ketoglutarate aminotransferase is of substantial physiological interest because it is involved in such diverse roles as amino acid metabolism, electron shuttling, intercellular transport of metabolites during C_4 photosynthesis, and glycolate metabolism. Six isozymes of this enzyme are present in the cotyledons of cucumber seedlings;

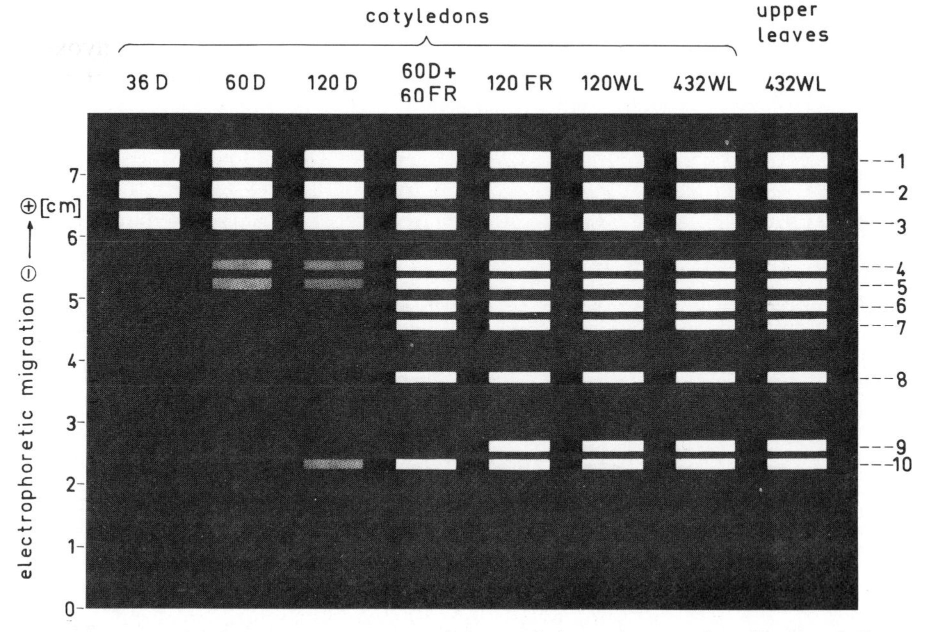

Fig. 6.4. Catalase isozyme patterns of the cotyledons of mustard seedlings as influenced by continuous far-red light (phytochrome activation) or white fluorescent light (7000 lux). Seedlings were grown either in darkness (D) for 36, 60, or 120 hr or under far-red (FR) or white light (WL) for 60 or 120 hr, respectively. For comparison, cotyledons or upper leaves from older (432 hr) greenhouse-grown plants were used. From Schopfer, *et al.* (1976b).

isozymes 1 and 2 are associated with the glyoxysomes, 3 in the cytosol, and 4, 5, and 6 are plastid isozymes (Liu and Huang, 1977). These isozymes, as might be expected, behave differently following germination. The activities of glyoxysomal and cytosolic isozymes increase until approximately day 2 and then begin to decrease (Fig. 6.5). The plastid isozymes appear later, independent of light treatment, and do not peak until about day 4. After peaking, the activity of glyoxysomal isozymes drop off rapidly, similar to the decrease in activity of other glyoxysomal enzymes in this tissue; the decrease is accelerated by light treatment. In the presence of light, activity of cytosolic isozyme also is reduced, but does not decrease to an undetectable level. From their peak, the activities of the plastid isozymes decrease about 30–40%, independent of light treatment. The temporal change in the isozyme pattern in the presence or absence of light parallels the development of other glyoxysomal enzymes and of chloroplasts, except that light has no effect on the appearance of the plastid isozymes.

C. Malate Dehydrogenase

Another enzyme that has received attention is malate dehydrogenase, also of special interest due to its multiplicity of roles and occurrence in three different cellular compartments, i.e., the cytosol, mitochondria, and peroxisomes (Ting *et al.*, 1975; see Chapter 4 for a more extensive discussion). In the cotyledons of cucumber seedlings, there are five isozymes: isozyme 1 is glyoxysomal or leaf-type peroxisomal, isozyme 3 is mitochondrial, and isozymes 2, 4, and 5 are cytosolic (Liu and Huang, 1976). There are distinct differences in the development of these isozymes over an 8-day period following seed imbibition (Fig. 6.6). The glyoxysomal activity is very low in the dry seed, but increases rapidly following imbibition until a peak is reached at day 3–4; a slow decline follows. Light has little effect other than accelerating the decline slightly. This lack of effect presumably results from the reciprocal appearance of leaf-type peroxisomes which contain an isozyme with characteristics very similar if not identical to those of the glyoxysomal isozyme. The glyoxysomal malate dehydrogenase is synthesized by cytoplasmic ribosomes (80 S) and not by mitochondrial or plastid ribosomes (70 S), as revealed by the effect of specific inhibitors applied *in vivo* (Walk and Hock, 1977). The subunit of the enzyme is synthesized initially as a precursor in the cytosol, and is larger than the final product in the glyoxysomes (Chapter 5).

In summary, those enzymes that are duplicated in different organelles presumably are coded by different genes and, in many cases, are regulated independently.

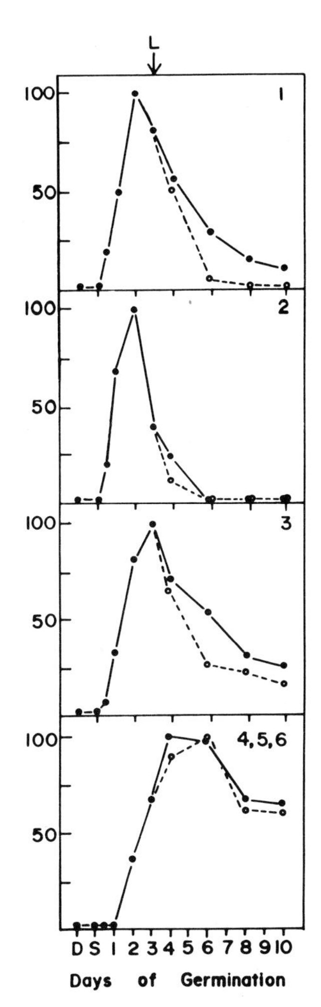

Fig. 6.5. Changes in aspartate–α-ketoglutarate aminotransferase isozyme activities in the cotyledons of cucumber seedlings. Soaked seeds were grown in darkness for 3 days following which the seedlings were either maintained in darkness (●) or transferred to continuous light (○). The amounts of each isozyme are expressed as a per cent of the peak stage of development. D and S are dry and soaked seeds, respectively. Isozymes 1 and 2 are glyoxysomal, 3 is cytoplasmic, and the remainder is found in plastids. From Liu and Huang (1977).

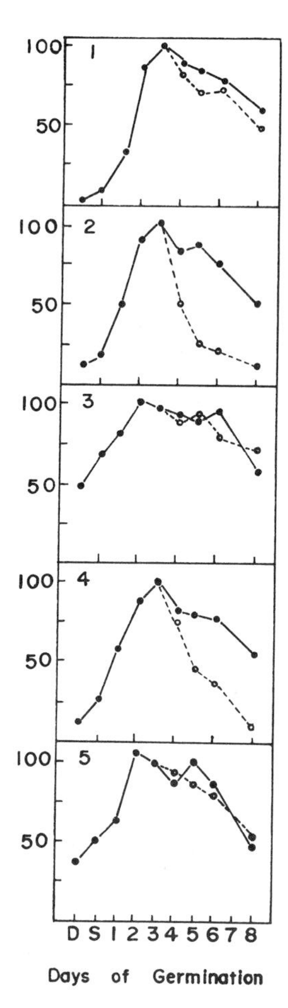

Fig. 6.6. Changes in malate dehydrogenase isozyme activities in cotyledons of cucumber seedlings. Seeds and seedlings were treated as those of Fig. 6.5 and the symbols are the same. Isozyme 1 is glyoxysomal or leaf-peroxisomal, 3 is mitochondrial, and the remainder cytosolic. From Liu and Huang (1976).

III. REGULATION

A. Light

Light has long been known to promote the development of peroxisomal enzyme activities in leaves and cotyledons of many species. In general, light stimulates the development of activities of peroxisomal enzymes in leaves and cotyledons, whereas in the cotyledons of oilseeds, it also accelerates the decrease of glyoxylate cycle enzyme activities. Similar effect of light on *Euglena* also has been observed (Chapter 4).

At least a portion of the peroxisomal response to light is controlled by phytochrome (Schopfer *et al.*, 1975, 1976b). Cotyledons of mustard (*Sinapis alba* L.) have been subjected to pulse illumination with red light and high-irradiance light treatment (continuous irradiation with far-red light of about 720 nm). The red light treatment shifts the phytochrome from P_r to the P_{fr} form, which is physiologically active. The high-irradiance treatment maintains a balance of P_r and P_{fr} capable of inducing phytochrome responses while minimizing effects due to photosynthesis. In general, enzyme activities of leaf-type peroxisomes, such as glycolate oxidase and hydroxypyruvate reductase, are stimulated by red or high-irradiance light treatment, while development of isocitrate lyase, a glyoxysomal enzyme, is not affected (Fig. 6.7). It is interesting that unlike the high-irradiance or red light effects, white light accelerates the decrease in glyoxysomal enzyme activities (Chapter 5). Far-red light treatments also affect the production of specific catalase isozymes as mentioned earlier in this chapter.

In the cotyledons of mustard seedlings, light may induce the formation of leaf-type peroxisomes as well as peroxisomes for ureide metabolism (Hong and Schopfer, 1981). During the growth of seedlings, activity of isocitrate lyase, the glyoxysomal marker, rises and falls as expected; light accelerates its decrease (Fig. 6.8). Treatment with white light induces the appearance of glycolate oxidase activity and chlorophyll. However, light through the phytochrome system also induces the appearance of urate oxidase activity, but its appearance lags several days behind that of glycolate oxidase. These observations were interpreted to indicate that three types of peroxisomes are formed in sequence in the cotyledons: glyoxysomes for gluconeogeneis, light-induced leaf-type peroxisomes for photorespiration, and light-induced but expression-delayed peroxisomes for urate oxidation of unknown physiological function.

Although phytochrome plays a role in the light stimulation of enzyme activities in leaf-type peroxisomes, it may not be the only photoreceptor involved. In wheat seedlings, blue, red, far-red, and white light stimulate the development of catalase, glycolate oxidase, and hydroxypyruvate reductase activities to different extents (Feierabend and Beevers, 1972a). Whereas the results of short-term light treatments indicate a role for the phytochrome system (stimulated by red light, reversed by far-red), the total response to this treatment is not as strong as that obtained

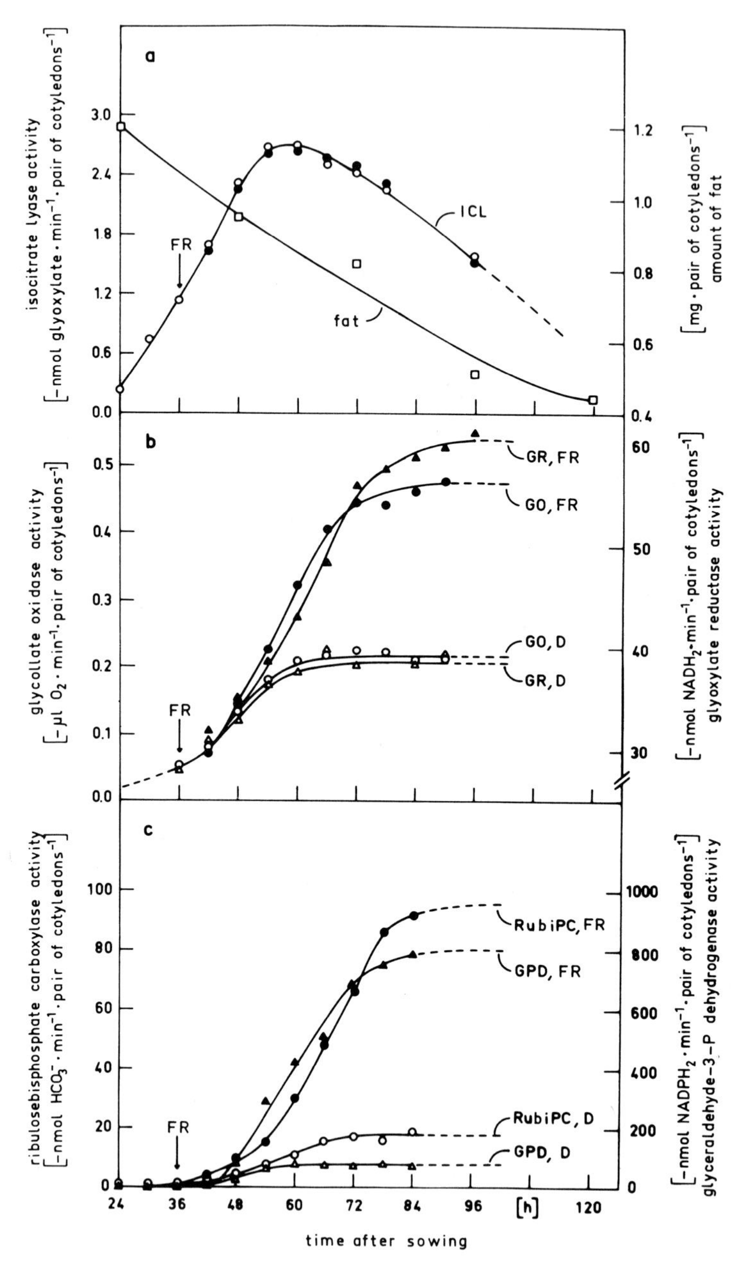

Fig. 6.7. Influence of continuous far-red light on the development of some glyoxysome (a), leaf-type peroxisome (b), and plastid (c) marker enzymes in the cotyledons of mustard seedling. Open circles are dark controls, and closed circles represent results in the presence of continuous, high-irradiance far-red light. From Schopfer *et al.* (1976b).

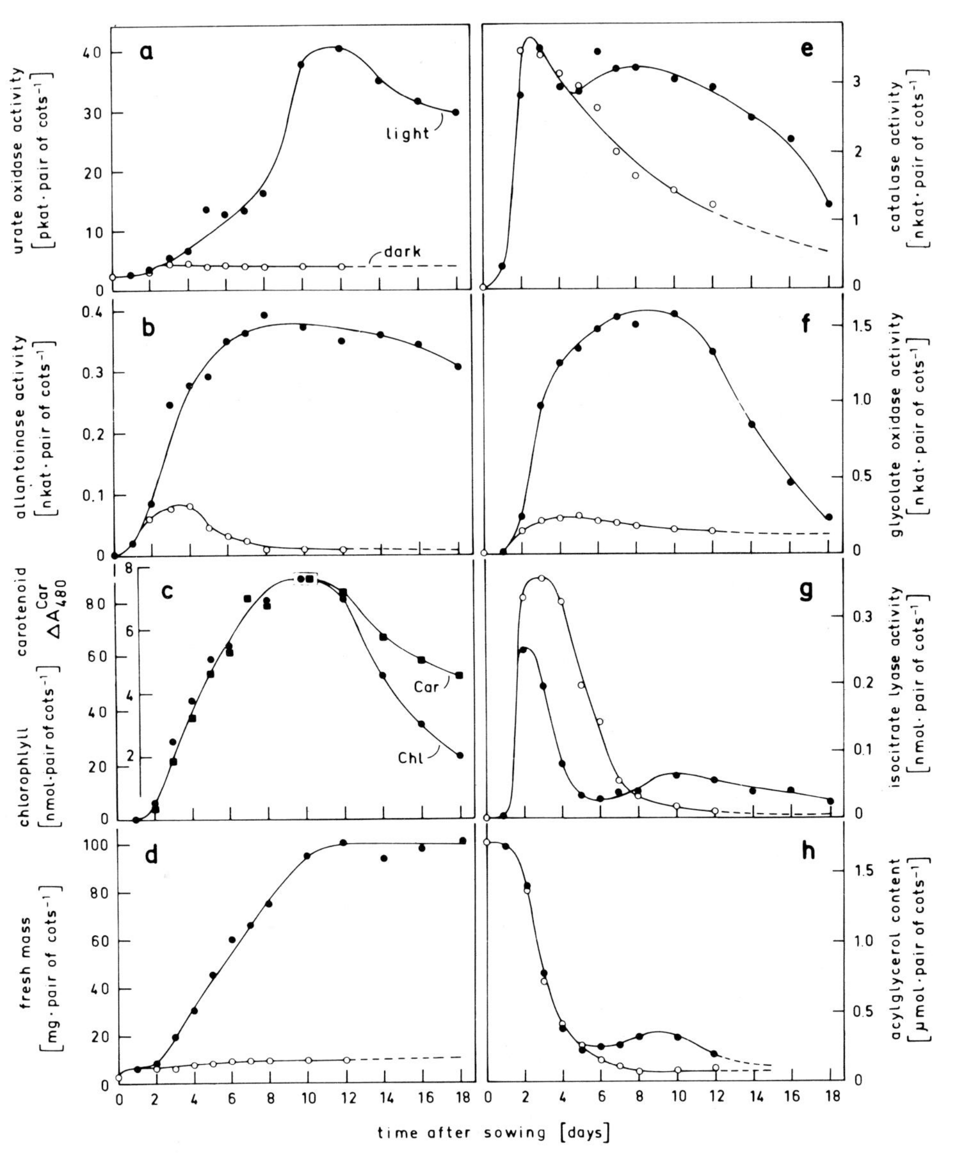

Fig. 6.8. Time course of various parameters related to peroxisomal functions in the cotyledons of white light-grown (●, ■) and dark-grown (○) mustard seedlings. From Hong and Schopfer (1981).

with continuous blue light of quantum-flux densities similar to that of red and far-red light. Such blue light treatment results in greater increases in catalase and glycolate oxidase activity (Fig. 6.9). This and related information suggest that some pigment system other than phytochrome also may be involved in the induction of enzymes in leaf-type peroxisomes.

Although there is a close functional relationship between chloroplasts and leaf-type peroxisomes such that a coordinated development of the two organelles generally occurs, the development of the two organelles being under separate control mechanisms could be demonstrated. In a few tissues, the development of some peroxisomal enzyme activities can occur in continuous darkness when chloroplast development is minimal or not evident. In mustard cotyledons, continuous far-red light treatment stimulates the development of enzyme activities in leaf-type peroxisomes but not chlorophyll (Schopfer *et al.*, 1976b). Similar light treat-

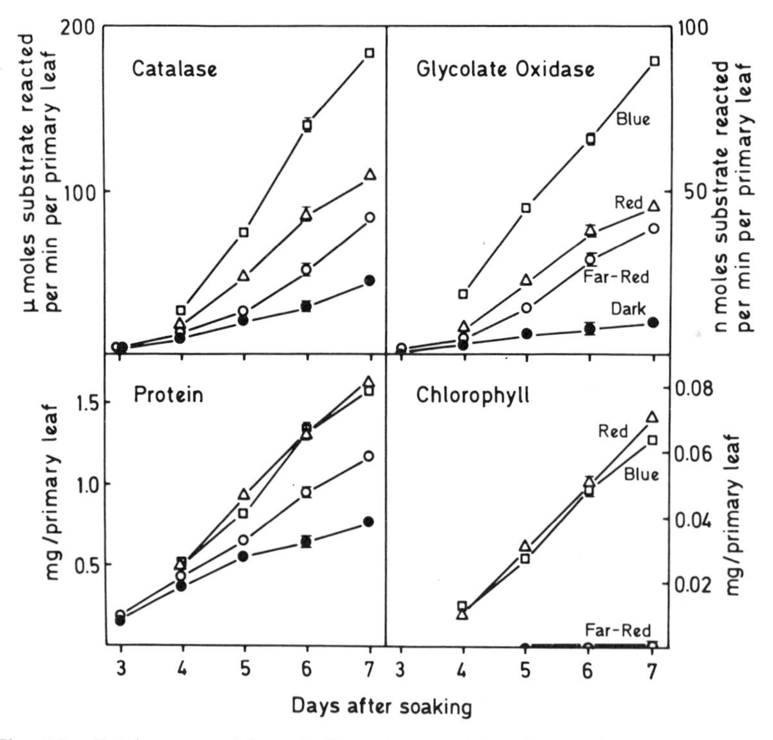

Fig. 6.9. Total protein, chlorophyll, catalase, and glycolate oxidase in extracts of first leaves of wheat seedlings growing in continuous darkness (●), or in far-red (○), red (△), or blue (□) light of similar quantum-flux density. From Feierabend (1975).

ment also induces an uncoordinated development of enzymes in peroxisomes and chloroplasts in wheat leaves, although some chloroplast development does proceed (Feierabend and Beevers, 1972a). Blue light stimulation of peroxisomal enzyme development in wheat leaves occurs even in the presence of aminotriazole at concentrations that inhibit chlorophyll formation some 80–90%. In rye, plastid protein synthesis activity is inhibited by growing the plants at 32°C, while cytoplasmic protein synthesis remains active. Peroxisomal enzymes continue to increase under these conditions, even though the plastids do not develop (Feierabend and Schrader-Reichardt, 1976). In *Euglena,* the Calvin cycle and the glycolate pathway are not coordinately regulated since, during the regreening of bleached cells, maximal glycolate dehydrogenase activity was reached before the complete development of RuBP carboxylase (Davis and Merrett, 1975).

B. Hormones

Hormones have been shown to stimulate germination and seedling development in a variety of species. The role of hormones in peroxisomal development has received only limited attention, and most of the results are inconclusive. Auxins, cytokinins, and gibberellins all cause an alteration in activities of certain peroxisomal enzymes, but the mechanism is unknown. It is likely that these hormones exhibit a more general influence on cell development or differentiation rather than a specific effect on peroxisome development.

The effects of gibberellins in castor bean endosperm have received some attention. Excised endosperm provided with water appears to go through the normal postgermination development independent of the embryonic axis or the cotyledons (Huang and Beevers, 1974). Treatment of the endosperm with 3×10^{-5} *M* GA_3 and GA_7 enhances development of the activity of isocitrate lyase activity on a fresh weight basis, and abscisic acid at a similar concentration counters this effect (Marriott and Northcote, 1977) (Fig. 6.10). Cyclic AMP also stimulates the activity when tested at the high concentration of 10^{-3} *M*. Exposure to 3×10^{-4} *M* GA_3 results in increased mitochondrial and glyoxysomal enzyme activities, but does not affect the total glyoxysomal or endoplasmic reticulum protein or the activity of choline phosphotransferase, an enzyme of the endoplasmic reticulum (Wrigley and Lord, 1977). The above findings suggest that gibberellin at these concentrations does not affect the assembly of glyoxysomes or the proliferation of the endoplasmic reticulum. On the other hand, distinct changes in these two organelles occur after a 24-hr treatment with gibberellin at lower concentrations

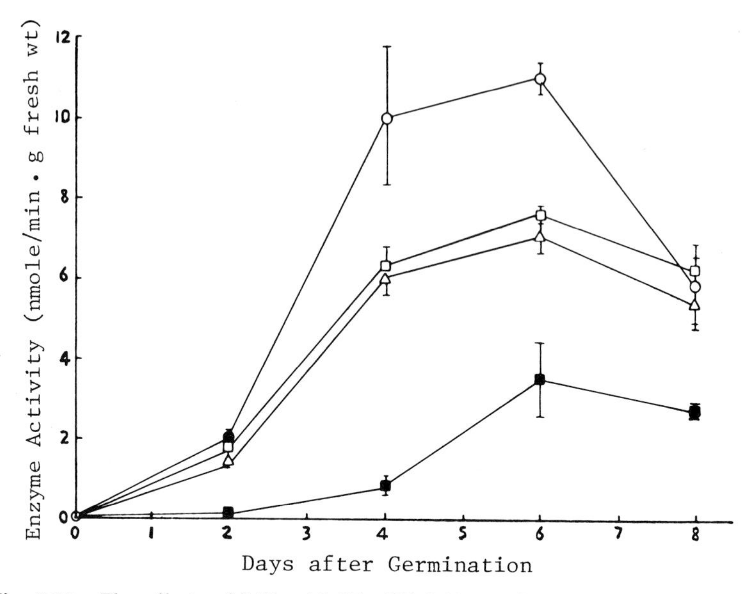

Fig. 6.10. The effects of 0.03 m*M* GA_3 (○) 0.03 m*M*abscisic acid (■), or equimolar amounts of both (△) on isocitrate lyase of the endosperm of castor bean in postgermination. Beans were germinated on agar gels containing water (□) or the hormone solutions. From Marriott and Northcote (1977).

(7×10^{-5} *M*) (Gonzalez, 1978). Two endoplasmic reticulum enzymes, NADH: cytochrome *c* reductase and phosphorycholine glyceride transferase, have increased activities in the presence of GA_3 (Fig. 6.11). Malate synthase found in the same region of the sucrose gradients as the endoplasmic reticulum fractions also has increased activity. In addition, the equilibrium density of the endoplasmic reticulum, as identified by NADH:cytochrome *c* reductase, increases in the sucrose gradients. Also, there is an increase in the specific activities of several enzymes of the glyoxysomal fractions in the sucrose gradients. In summary, castor bean endosperm can go through normal postgermination development by itself without the embryonic axis or cotyledons. Externally applied gibberellins at rather high concentrations apparently enhance its development, and the physiological role of gibberellins under normal physiological conditions is unclear.

Effects of kinetins on peroxisome development have been reported, but again the physiological significance of these effects is unknown. None of the observed kinetins effects has been shown to be specific for peroxisomal development, and thus the hormonal effects may be only on general cell development and differentiation. Moreover, these investigations utilize excised cotyledons of oil seedlings, and thus the influence of removal of sink (embryonic axis) on the supply organ (cotyledons) is an additional unknown factor. In general, derooting (the embryonic axis) of the cotyledons causes a decline in lipid mobilization and glyoxylate cycle enzyme activities, as well as an accumulation of soluble sugars and amino acids. Application of kinetins at about 10^{-5} *M* overcomes this decline or hastens the increase or decrease (depending on the developmental stage of the cotyledons) of glyoxylate cycle enzyme activities and the increase of leaf-type peroxisomal and chloroplast enzyme activities. Thus, kinetins promote the conversion of the cotyledons from a storage to a photosynthetic function. This kinetins effect has been observed in the cotyledons of sunflower (Servettaz *et al.*, 1976; Theimer *et al.*, 1976), squash (Penner and Ashton, 1967), and watermelon (Longo *et al.*, 1979). The enhancement of development of photosynthetic machinery by kinetins also occurs in derooted rye seedlings grown in darkness (DeBoer and Feierabend, 1974).

C. Metabolites

The role of metabolites and other small molecules in regulating the formation, destruction, and activities of various organelles and their

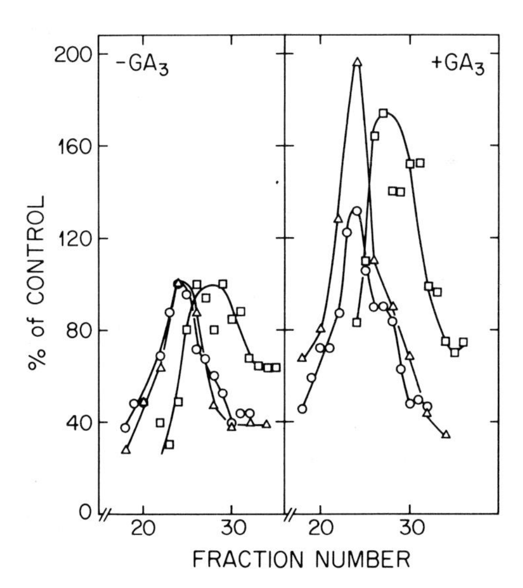

Fig. 6.11. Relative values of malate synthase (□), NADH-cytochrome *c* reductase (○), and phosphorylcholine–glyceride transferase (△) at the region of the endoplasmic reticulum fractions in sucrose gradients obtained from GA_3-treated (0.07 m*M*) and untreated castor bean endosperm. From Gonzalez (1978).

enzymes in diverse organisms has been studied extensively. Such regulation might be expected for peroxisomes. Indeed, numerous metabolites have been implicated in such controls. The induction of peroxisomal enzymes as well as peroxisome proliferation by alkanes, fatty acids, fatty alcohol, acetate, ethanol, methanol, and amines in microorganisms have been described in Chapter 4. Less information is available in higher plants. In suspension cultures of *Anise,* external supplies of acetate or ethanol induce an increase in the activities of glyoxylate cycle enzymes and the number of glyoxysomes (Kudielka *et al.,* 1981). Similarly, alteration of glyoxysomal enzyme development with added sugars, fatty acids, glutamine, and hydroxylamine has been reported in many tissues (Gerhardt, 1978), but the physiological significance of these findings is not clear. When roots of *Phaseolus coccineus* were grown in elevated nitrogen salts, activities of urate oxidase and catalase in crude extract and in isolated peroxisomes increased (Theimer and Heidinger, 1974). The implication is that the nitrogen salts are assimilated into purines, which are metabolized by urate oxidase in the peroxisomes to produce allantoin, the transport form of nitrogen in the bean plants. At the level of enzyme molecules, direct modification of the *in vitro* activities of numerous peroxisomal enzymes from diverse species by various metabolites has been demonstrated (reviewed in Gerhardt, 1978).

D. Inhibitors

Although the role of inhibitors in controlling the activities of a variety of enzymes in diverse species has been well documented, studies on specific inhibitors of peroxisomal enzymes are few and incomplete.

An inhibitor of catalase is present in maize seedlings and appears at specific stages of development (Sorenson and Scandalios, 1980). The inhibitor occurs in high amounts in the aleurone layer and scutellum, but in low amounts in other tissues of the seedlings. This distribution correlates well with the specific activity of catalase in the tissue extracts. The inhibitor is not effective against peroxidase or a few other enzymes, but exerts inhibition on catalase from diverse species. The inhibitor inhibits 20–40% of the catalase activity, and its subcellular location is unknown.

Proteinaceous inhibitors of isocitrate lyase activity have been reported from spinach leaves (Godovari *et al.,* 1973), sunflower cotyledons (Theimer, 1976), and preclimacteric bananas (Surendranathan and Nair, 1978). The inhibitor was purified 160-fold from banana, and found to contain glutamate and proline as the major amino acids.

The precise role of these inhibitors is not known. They may be a part of the mechanism for regulating enzyme activities at particular stages of development.

E. Coordinate Enzyme Development

Many studies on the development of enzyme activities following seed germination and greening of leaves revealed that there is a parallel development of activities of various peroxisomal enzymes (e.g., Fig. 5.12). These observation plus the idea of the ER-vesiculation model of peroxisome biogenesis contributed to the generalization that peroxisomal enzymes are coordinately synthesized. Recent work with maturing seeds indicate that to a certain extent, the glyoxysomal enzymes are not synthesized in a totally synchronous fashion (Chapter 5, Section III,B). Furthermore, the coordinate development of the various enzyme activities can be altered in genetic mutants or by artificial manipulations (e.g., light treatments), suggesting that the formation of these enzymes is controlled by separate mechanisms.

In the endosperm of germinated castor beans, the increase in malate synthase activity precedes that of catalase (Gonzalez, 1978). In cotton and cucumber seeds during maturation, about 20% of some glyoxysomal enzyme activities are present, and the remaining 80% of these enzyme activities together with 100% of the activities of other glyoxysomal enzymes appear after germination (Chapter 5, Section III,B). In a mutant maize line (A 16), catalase activity in the scutellum of germinated seeds is very low due to the lack of expression of the *Cat 2* gene (Fig. 6.12). However, the levels of isocitrate lyase and malate synthase activities remain high, as in normal corn (Tsaftaris and Scandalios, 1981). During the formation of leaf peroxisomes in illuminated wheat leaves, the appearance of catalase activity exhibits a longer lag period than other peroxisomal enzyme activities (Fig. 5.16) (Feierabend and Beevers, 1972a). The wheat catalase also responds differently to far-red light stimulation than other peroxisomal enzymes (Feierabend, 1975).

IV. SUMMARY

Of all peroxisomal enzymes in higher plants, only catalase in maize is known to be coded by nuclear genes whose expression is tissue-specific and controlled by regulatory genes. Peroxisomal aspartate–α-ketoglutarate aminotransferase and malate dehydrogenase exist as distinct subcellular isozymes; during postgerminative seedling growth or leaf greening, each isozyme exhibits a temporal developmental change in activity

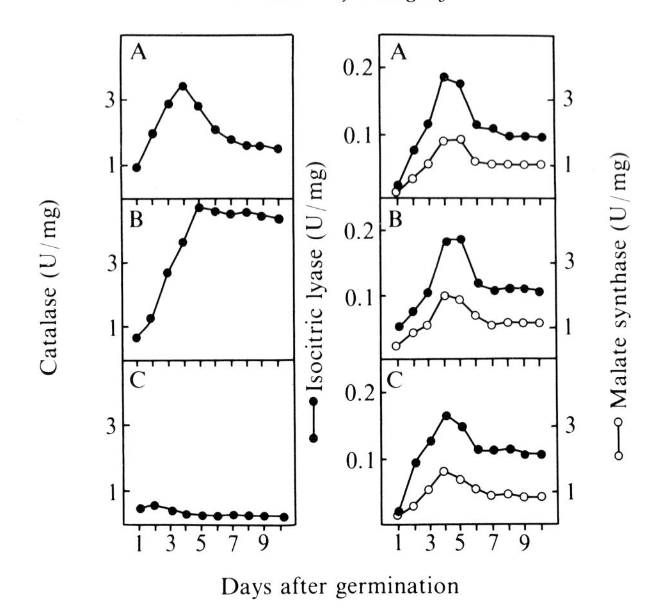

Fig. 6.12. Activities of catalase, isocitrate lyase, and malate synthase in scutellar extracts of W64A (A), R6-67 (B), and A16 (C) in postgermination. Redrawn from Tsaftaris and Scandalios (1981).

that is different from that of the other isozymes. Phytochrome is involved in regulating the appearance of some of these isozymes and leaf peroxisomes in general. Other light receptors also may be involved. Effects of phytohormones, metabolites, and inhibitors on the development of peroxisomal enzyme activities also have been demonstrated. Increases in peroxisomal enzyme activities during development appear to be coordinated, but noncoordination, indicative of these enzymes being controlled by separate mechanisms, has been shown in a few cases occurring naturally or by artificial manipulation.

7

Perspective

I. INTRODUCTION

Despite the recent advances in our understanding of peroxisomal structure and function, it is our opinion that some of the most exciting research remains to be done. Full elucidation of the structure of these organelles, their functions in specialized tissues, their ontogeny, and the regulation of genetic expression of peroxisomes need to be actively pursued. Further research on peroxisomes should be very rewarding.

II. NOMENCLATURE

In this monograph, we have adopted a consistent use of the term peroxisome for all the related organelles that have been shown to possess similar peroxisomal characteristics. These characteristics include the morphology, and the content of one or more of the following enzymes: (1) catalase, (2) H_2O_2-producing oxidases, (3) a complete or an incomplete complement of glyoxylate cycle enzymes, and (4) glycolate metabolizing enzymes. Up to the present time, the term *microbody* has been used commonly to describe these organelles in plant cells. It is clear to us from the information presented in this monograph that most, if not all, plant microbodies are types of peroxisomes. We urge plant biologists to abandon using microbodies to name these organelles, as has already been done by mammalian and microbial biologists. The basic concept of peroxisomal function should not be obscured by variation in terminology. Glyoxysomes are a type of peroxisomes, and the two terms should not be used equally in the nomenclature hierarchy. The term *glyoxysome* should be retained to describe the specialized peroxisomes in oilseeds that contain a complete complement of glyoxylate cycle enzymes.

The terminology as described in the preceding paragraph is appropriate for organelles in higher plants. It is also suitable for organelles in primitive or unicellular eukaryotic organisms, even though several special cases deserve additional comments. In fungi, glyoxysome-like organelles do not possess a complete complement of glyoxylate cycle enzymes, and some may deem it inappropriate to call these fungal organelles strictly glyoxysomes. In this monograph, we call them peroxisomes or "glyoxysome-like peroxisomes." In *Neurospora,* the glyoxysome-like organelles may be devoid of catalase and distinct from other peroxisomes in the same cell. In *Euglena,* the glyoxysomes and the leaf-peroxisome-like organelles may contain no catalase and thus lack a basic characteristic of a peroxisome. However, these observations on *Neurospora* and *Euglena* require confirmation (Chapter 4).

III. MORPHOLOGICAL AND CYTOCHEMICAL STUDIES

Electron microscopic studies have established the ubiquitous occurrence of peroxisomes in diverse cell types within the plant kingdom. The shape, size, number per cell, and relationship to other organelles are extremely variable. This variation reflects diversity in cell and tissue types, taxonomic origins, as well as metabolic functions of the organelles. Although morphological studies may be considered classical in view of the rapid advances in molecular biology frontiers, they nevertheless serve an indispensable role and still provide valuable information. For example, the recent discovery of peroxisome enlargement and proliferation in cells adjacent to the *Rhizobium*-infected cells in root nodules of certain legume species was made from morphological studies and could not have been achieved by biochemical means. It is likely that peroxisomes functionally important in yet undiscovered metabolic processes occur in many specialized, unobstrusive tissues; morphological studies could pave the way toward discovering these peroxisomes.

Counting the number of peroxisomes per cell and serial sectioning can and need to be done microscopically. These approaches will provide a better understanding of the relationship between peroxisomes and other cellular compartments, and give much insight into the origin of peroxisomes.

Cytochemistry is an invaluable adjunct to ultrastructural studies, and has played a significant role in peroxisomal research. The cytochemical localization of a specific H_2O_2-producing oxidase can help define the function of peroxisomes observed *in situ,* and delineate the development of such a function within the organelle. Another significant application would be determining whether all of the peroxisomes within a cell shared the same enzyme complement; such an *in situ* evaluation would be more direct than the *in vitro* biochemical studies of subcellular fractions. To further exploit the *in situ* study of peroxisomes, the development and application of immunocytochemical procedure are valuable. These methods are being developed for study of other organelles and membrane systems, and would have a substantial impact on peroxisome research.

IV. MOLECULAR STRUCTURE OF PEROXISOMES

We have limited knowledge of the organization of proteins and lipids in peroxisomes. Glyoxysomes are known to have enzymes associated with the membranes, but the details of this association are lacking. Knowledge of the arrangement of enzymes, particularly their sidedness and proximity to other enzymes, in the glyoxysomal membrane is essential for a complete understanding of metabolic flow during active metabolism and of assembly mechanisms during organelle biogenesis. In leaf peroxisomes, none of the enzymes examined so far is associated with the membrane. In view of the sequential reactions occurring in leaf peroxisomes, it seems quite unlikely that the enzymes in leaf (and other) peroxisomes are randomly packed. It is more likely that the enzymes, such as catalase and the H_2O_2-producing oxidases, are aggregated in order to ensure a defined flow of metabolites. Such an idea is not without precedent; the loose association of glycolytic enzymes in the cytosol and of malate dehydrogenase and aspartate-α-ketoglutarate aminotransferase in the mitochondria of mammalian systems has been documented.

V. METABOLISM

Peroxisomes have been viewed as organelles engaged in a primitive form of respiration. Unlike mitochondrial oxidation, peroxisomal oxidation does not generate useful metabolic energy in the form of ATP. Thermogenesis has been proposed to be a function of peroxisomal respiration in mammalian adipose tissues (Section VII); this function does not appear to be important in higher plants. The need of plant peroxisomes as distinct respiratory organelles can be conceptualized in two ways. The first is the necessity for compartmentation of H_2O_2-producing oxidases and catalase together, such that the H_2O_2 produced is rapidly decomposed by catalase before it can exert toxic effect on cellular constituents. The second is the need for confining together the enzymes that are used in the production and consumption of glyoxylate. Such design would prevent the highly reactive glyoxylate from participating in undesirable reactions. All the peroxisome types fall into either or both of these concepts.

In higher plants, specialized peroxisomes now include glyoxysomes, leaf peroxisomes, and peroxisomes for ureide metabolism. It seems possible that additional specialized peroxisomes will be discovered in the future. There are still many conditions under which peroxisomal metabolism has not been examined. Examples are peroxisomes in certain tissues at different stages of development or differentiation. Another potentially important area of study is the peroxisomes in tissues or organs under various environmental stress conditions. Also, if we examine the list of H_2O_2-producing oxidases present in eukaryotes, it is evident that the subcellular location of many of these oxidases has not been examined and, in some cases, are presumed to be mitochondrial (data from crude particulate fractions). The possible localization of these enzymes in the peroxisomes should be studied.

Procedures to isolate intact peroxisomes in low osmotic-strength solution have been developed very recently (see Chapter 3). These peroxisome preparations should be used in the study of membrane permeability and organelle metabolism under physiological conditions. Also, they should be used in elucidating the mechanism of organelle biogenesis such as when testing whether the intact organelles can process and acquire a specific enzyme synthesized in an *in vitro* translation system (Chapter 5).

The varying compartmentation of peroxisomal enzymes in higher plants and eukaryotic microorganisms is of interest from an evolutionary viewpoint. In fungi, as well as in *Tetrahymena,* the glyoxysome-like peroxisomes contain only malate synthase and isocitrate lyase, and the other three glyoxylate cycle enzymes are present in the mitochondria. In contrast, the glyoxysomes in oilseeds of higher plants and the glyoxysome-like peroxisomes in *Euglena* contain all five enzymes of the glyoxylate cycle. Glyoxysomes apparently are functional in spores of mosses and ferns in postgermination, but information is not available to tell whether they have a complete complement of glyoxylate cycle enzymes. It is possible that the genes for the three extra enzymes (aconitase, citrate synthase, and malate dehydrogenase) are originated and duplicated from the nuclear genes of the mitochondrial TCA-cycle enzymes, and were added to the peroxisomal control system late in evolution.

Another aspect of evolutionary interest is the occurrence of glycolate dehydrogenase in most algae and glycolate oxidase in other algae and all higher plants. In *Euglena*, and perhaps also in other algal species that contain glycolate dehydrogenase, the enzyme is present in both peroxisomes and mitochondria at approximately equal activity. The peroxisomal enzyme transfers electron to a still-unknown electron acceptor, but not to molecular oxygen. In contrast, the mitochondrial enzyme transfers electrons to the electron transport chain and thus useful metabolic energy is generated. In this aspect, the mitochondrial enzymic reaction is more beneficial to the cellular energetics than that of the peroxisomal enzyme. However, it is the peroxisomal enzyme, and not the mitochondrial enzyme, that ostensibly is active in glycolate oxidation during photorespiration; it is unclear why the mitochondrial enzyme that can generate useful metabolic energy is not involved. Since glycolate dehydrogenase is also present in blue green algae and bacteria, it should be considered evolutionarily more primitive than glycolate oxidase of some algae and all higher plants. The advantage of having glycolate oxidase over glycolate dehydrogenase is unclear, especially since glycolate oxidase transfers electrons directly to molecular oxygen without the production of useful metabolic energy. Perhaps during evolution, glycolate dehydrogenase was transformed to glycolate oxidase, which became much more active on a per protein molecule basis. Thus, organisms with glycolate oxidase would have a higher capacity to oxidize glycolate, as is the case in higher plants in comparison with organisms containing glycolate dehydrogenase. The identification of the native electron acceptor of the glycolate dehydrogenase reaction is important to our understanding of glycolate metabolism as well as the evolution of the glycolate oxidizing enzymes.

VI. BIOGENESIS

Recent investigations suggest that the mechanism of peroxisome biogenesis does not follow closely the ER-vesiculation model, which emphasizes the vesiculation of complete organelles from specialized regions of the endoplasmic reticulum. It now appears likely that the membrane phospholipids and integral proteins are synthesized on the ER, whereas the membrane peripheral proteins and the matrix proteins are synthesized on unbound cytosolic polysomes. Where, when, and how these components are assembled are unknown. The peroxisomes may be quite capable of acquiring new enzymes and, perhaps, deleting old ones. With several laboratories actively pursuing these problems, the mechanism should become much clearer within a few years.

The elucidation of peroxisome biogenesis in higher plants may be hindered by the nature of the study systems. Tissues such as endosperm or organs such as cotyledons are not composed of cells at the same developmental stage at a particular time. Moreover, the cotyledon is composed of epidermal cells, vascular bundles, and guard cells, in addition to the majority of mesophyll parenchyma cells. These characteristics indicate that data obtained from cell fractionation and accompanying radio-tracer studies are some "average" of the changing events and perhaps of the different types of peroxisomes in these tissues and organs. Ultrastructural studies are hampered by being able to observe only a very thin section of the large cells, and therefore only an extremely small portion of the whole tissue. However, coupling cell fractionation, light and electron microscopy, cytochemistry, and tracer experiments can and have been done to make inroads into understanding the mechanism of peroxisome biogenesis. Also, more innovative approaches to the problem likely will be developed in future studies.

Unicellular fungal cells, especially *Candida*, offer some technical advantages in the study of peroxisome biogenesis. The *Candida* cells are small, containing about four to six peroxisomes per cell. Counting the number of organelles per cell during a period of biogenesis with significant statistical values can be achieved. Serial sectioning of the whole cell is easily accomplished, so that if vesiculation and division of peroxisomes occurs, it can be observed. Specific changes in the metabolic role of the peroxisomes can be induced by simply altering the carbon or nitrogen source in the media. So far, peroxisomal biogenesis in these unicellular fungi has been studied mostly by electron microscopy. We suggest that the fungal system should be explored in greater detail with respect to the mechanism of peroxisomal biogenesis. However, we should bear in mind that the fungal system is different from that of higher plants, and therefore the research on higher-plant peroxisome biogenesis should be pursued concurrently.

Traditionally, a series of comprehensive studies on an organelle proceeds from its morphology to its physiology/metabolism, to its ontogeny, and then to the genes involved and to their regulation. The studies on peroxisomes have not been an exception. Major advances have been made on the morphology and physiology/metabolism of peroxisomes in the past 15 years. With active research efforts currently being devoted to understanding peroxisome ontogeny, the mechanism of biogenesis should become clearer within the next several years. Perhaps it is time that we should start looking ahead to investigating the genes involved in regulating peroxisome function and biogenesis. To date, we know that in maize there are several structural genes and at least one regulatory gene of catalase that are under temporal control. We also know that several peroxisomal enzymes are synthesized on 80 S rather than 70 S ribosomes. Studies of gene regulation at the molecular biology level are the new frontiers that need to be pursued actively. These new studies should not be initiated at the expense of terminating research on other aspects of peroxisomes. There are still major gaps in our knowledge that need to be resolved from morphological, physiological, metabolic, and ontogenic studies as have been pointed out throughout the chapters.

VII. PEROXISOMES IN NONPLANT TISSUES

Although the peroxisomes from liver were the first peroxisomes to be isolated, their roles in metabolism have been more explicitly defined only recently (see reviews by de Duve and Baudhuin, 1966; Lazarow, 1981; Masters and Holmes, 1977; Fahimi and Yokota, 1981; Tolbert, 1981). Earlier, the liver peroxisomes were shown to oxidize ureide by urate oxidase, allantoinase and allantoicase, D-amino acid by D-amino acid oxidase, and ethanol by the peroxidative activity of catalase, but the relevance of these reactions to the overall metabolism of the cell is still unclear. Recently, capacity for β-oxidation of fatty acid and its activation were discovered in liver peroxisomes. Unlike the oilseed tissues in which the glyoxysomes are the exclusive subcellular site of fatty acid oxidation, mammalian liver possesses these activities both in the peroxisomes and in the mitochondria. The peroxisomal fatty acyl-CoA oxidase donates electrons directly to molecular oxygen, whereas the mitochondrial fatty acyl-CoA dehydrogenase passes electrons to the electron transport chain. The peroxisomal system oxidizes fatty acids of chain length longer than eight carbon atoms; the oxidation of shorter-chain fatty acids produced in the peroxisomes apparently is carried out by the mitochondria in the same cell. The peroxisomes contain carnitine acyltransferase for the export of short-chain fatty acids and glycerol phosphate dehydrogenase for the export of NADH generated in β-oxidation. Under normal physiological conditions, the capacity of β-oxidation in peroxisomes in 10–50% of that in mitochondria (Lazarow, 1981; Neat *et al,* 1981). The activity of the peroxisomal, but not the mitochondrial oxidative system, is elevated, and the peroxisomes proliferate under conditions that favor lipid utilization. These conditions include a high-fat diet, administration of hypolipidemic drugs, and genetically obese mice. In addition to being found in liver, peroxisomal β-oxidation activity has also been found in other mammalian tissues and organs. An elevation of peroxisomes and their enzyme activities can be observed in adipose tissue when the animal is exposed to cold. In this last aspect, a role of peroxisomes in thermogenesis has been proposed in view of the fact that energy generated in peroxisomal β-oxidation of fatty acids is not conserved in the form of ATP but is released as heat.

Dihydroxyacetone phosphate acetyltransferase, an initial enzyme in glycerolipid metabolism, and flavin-linked polyamine oxidase also are present in rat liver peroxisomes. The role of these two enzymes has not been clarified.

According to some recent interpretations of biochemical and morphological data, the peroxisomes in rat liver are not connected directly to the endoplasmic reticulum. However, they apparently are interconnected, either transiently or permanently, in clusters through a smooth membrane system. The term "peroxisome reticulum" has been used to describe the clusters. Peroxisomes apparently originate and proliferate from this reticulum in response to certain stimuli (undefined as yet). Addition of matrix enzymes to "budding" peroxisomes of the reticulum appears to be by posttranslational insertion (Lazarow *et al,* 1982) as occurs in oilseeds (Chapter 5). Membrane components probably come from the peroxisome reticulum. Such a reticulum has not been observed in plant cells, and thus the source of membrane components may be different for plant and animal peroxisomes.

The glyoxylate cycle is known to be operative in protozoa and a few metazoan tissues (review by Cioni *et al,* 1981). In *Tetrahymena* and a few other protozoa (Müller, 1975), isolated peroxisomes contain malate synthase and isocitrate lyase, whereas the other three glyoxylate cycle enzymes are present in the mitochondria. The protozoan peroxisomes thus are similar to the fungal peroxisomes in having an incomplete complement of glyoxylate cycle enzymes. In some Nematodes, such as in *Ascaris* eggs during postfertilization development and in *Turbatrix* and *Caenorhabditis* adults grown on acetate, the glyoxylate cycle is a main pathway for utilization of the acetate. In *Ascaris* and *Turbatrix,* the localization of glyoxylate cycle enzymes exclusively in the mitochondria has been reported (McKinley and Trelease, 1980). However, in *Caenorhabditis,* glyoxysome-like organelles, which contain catalase, isocitrate lyase, and malate synthase and exhibiting an equilibrium density of 1.25 g/cm^3 in a sucrose gradient, have also been reported (Patel and McFadden, 1977). In toad urinary bladder epithelium, cyanide-insensitive fatty acyl-CoA oxidation (similar to the liver peroxisomal activity) as well as malate synthase and isocitrate lyase activities are present, suggesting the presence of glyoxysome-like organelles (Goodman *et al*, 1980). The peroxisomes have not been isolated from this tissue, but were stained *in situ* for malate synthase reactivity. Activities of glyoxylate cycle enzymes also have been reported recently in chick embryo and fetal guinea pig liver, but no work has been done on the organelles housing these enzyme activities. Until these reports, it was thought that glyoxysome-like organelles did not occur in vertebrate tissues. It is quite possible that glyoxylate cycle activity is present in some other vertebrate tissues but is expressed only at a certain stage of the life cycle. Whether or not the enzymes are localized in peroxisomes (glyoxysomes) remains to be seen. These studies will provide additional information on the evolution of the glyoxylate cycle and glyoxysomes.

Bibliography

Not all references on plant peroxisomes have been cited. There are more than 1500 related to plant peroxisomes, and it is not practical to cite all of them. The more recent or classical reviews include the following.

Plant peroxisomes: Frederick *et al.* (1975); Gerhardt (1978; a monograph written in German); Beevers (1979); Lord (1980); Tolbert (1980).

Microbial peroxisomes: Müller (1975; protozoa); Maxwell *et al.* (1977; phytopathogenic fungi); Fukui and Tanada (1979; yeast); Veenhuis *et al.*, 1982. (yeast on one-carbon compounds.)

Peroxisomes in general with emphasis on mammalian systems: de Duve and Baudhuin (1966; classic); Master and Holms (1977); Bock *et al.*, (1980); Lazarow (1981); Fahimi and Yokota (1981); Tolbert (1981), Tolbert and Essner (1981).

Ainsworth, J. C. (1973). Introduction and keys to higher taxa. *In* "The Fungi, an Advanced Treatise" (G. C. Ainsworth, F. K. Sparrow, and A. S. Sussman, eds.), Vol. IVA, pp. 1–7. Academic Press, New York.

Akhtar, M. W., Parveen, H., Kausar, S., and Chughtai, M. I. D. (1975). Lipase activity in plant seeds. *Pak. J. Biochem.* **8,** 77–82.

Alexopoulos, C. J. (1962). "Introductory Mycology," 2nd ed. Wiley, New York.

Anderson, N. G. (1966). Zonal centrifuges and other separation systems. *Science* **154,** 103–112.

Armentrout, V. N., and Maxwell, D. P. (1981). A glyoxysomal role for microbodies in germinating conidia of *Botryodiplodia theobromae*. *Exp. Mycol.* **5,** 295–309.

Armentrout, V. N., Graves, L. B., Jr., and Maxwell, D. P. (1978). Localization of enzymes of oxalate biosynthesis in microbodies of *Sclerotium rolfsii*. *Phytopathology* **68,** 1597–1599.

Bateman, D. F., and Beer, S. V. (1965). Simultaneous production and synergistic action of oxalic acid and polygalacturonase during pathogenesis by *Sclerotium rolfsii*. *Phytopathology* **55,** 204–211.

Baudhuin, P., Beaufay, H., and de Duve, C. (1965). Combined biochemical and morphological study of particulate fractions from rat liver. *J. Cell Biol.* **26,** 219–243.

Baudhuin, P., Evrard, P., and Berthet, J. (1967). Electron microscopic examination of subcellular fractions. *J. Cell Biol.* **32,** 181–191.

Becker, W. M., Leaver, C. J., Weir, E. M., and Riezman, H. (1978). Regulation of glyoxysomal enzymes during germination of cucumber. I. Developmental changes in cotyledonary protein, RNA, and enzyme activities during germination. *Plant Physiol.* **62,** 542–549.

Becker, W. M., Reizman, H., Weir, E. M., Titus, D. E., and Leaver, C. J. (1982). *In vitro* synthesis and compartmentalization of glyoxysomal enzymes in cucumber. *Ann. N. Y. Acad. Sci.* **386,** 329–348.

Beevers, H. (1956). Utilization of glycerol in the tissues of the germinating castor bean. *Plant Physiol.* **31,** 440–445.

Beevers, H. (1961). "Respiratory Metabolism in Plants." Harper and Row, New York.

Beevers, H. (1979). Microbodies in higher plants. *Annu. Rev. Plant Physiol.* **30,** 159–197.

Beevers, H. (1980). The role of the glyoxylate cycle. *In* "The Biology of Plants" (P. K. Stumpf and E. E. Conn, eds.), Vol. 4, pp. 117–130. Academic Press, New York.

Beevers, H., and Breidenbach, R. W. (1974). Glyoxysomes. *In* "Methods in Enzymology" (S. Fleischer and L. Packer, eds.), Vol. 31, pp. 565–571. Academic Press, New York.

Beevers, L., and Hageman, R. H. (1980). Nitrate and nitrite reduction. *In* "The Biology of Plants" (P. K. Stumpf and E. E. Conn, eds.), Vol. 5, pp. 115–168. Academic Press, New York.

Beezley, B., Gruber, P. J., and Frederick, S. E. (1976). Cytochemical localization of glycolate dehydrogenase in mitochondria of Chlamydomonas. *Plant Physiol.* **58,** 315–319.

Begin-Heick, N. (1973). The localization of enzymes of intermediary metabolism in *Astasia* and *Euglena*. *Biochem. J.* **134,** 607–616.

Bentley, R. (1963). Glucose oxidase. *In* (Boyer, P. D., Hardy, H., Myrbäck, K., eds.), "The Enzymes," 2nd edn, Vol. 7, pp. 567–586. Academic Press, New York.

Bergfeld, R., Hong, Y. N., Kuhne, T., and Schopfer, P. (1978). Formation of oleosomes (storage lipid bodies) during embryogenesis and their breakdown during seedling development in cotyledons of *Sinapis alba* L. *Planta* **143,** 297–307.

Bergner, U., and Tanner, W. (1981). Occurrence of several glycoproteins in glyoxysomal membranes of castor beans. *FEBS Lett.* **131,** 68–72.

Betsche, T., and Gerhardt, B. (1978). Apparent catalase synthesis in sunflower cotyledons during the change in microbody function. A mathematical approach for the quantitative evaluation of density-labeling data. *Plant Physiol.* **62,** 590–597.

Bieglmayer, C., and Ruis, H. (1974). Protein composition of the glyoxysomal membrane. *FEBS Lett.* **47,** 53–55.

Bieglmayer, C., Graf, J., and Ruis, H. (1973). Membranes of glyoxysomes from castor bean endosperm. Enzymes bound to purified membrane preparations. *Eur. J. Biochem.* **37,** 379–389.

Blobel, G., and Dobberstein, B. (1975). Transfer of proteins across membranes. 1. Presence of proteolytically processed and unprocessed nascent immunoglobulin light chains on membrane-bound ribosomes of murine myeloma. *J. Cell Biol.* **67,** 835–851.

Bock, P., Kramar, R., and Pavelka, M. (1980). Peroxisomes and related particles in animal tissues. Springer-Verlag, New York.

Bold, H. C., and Wynne, M. J. (1978). "Introduction to the Algae." Prentice-Hall, Englewood Cliffs, New Jersey.

Bortman, S. J., Trelease, R. N., and Miernyk, J. A. (1981). Enzyme development and glyoxysome characterization in cotyledons of cotton seeds. *Plant Physiol.* **68,** 82–87.

Bowden, L., and Lord, J. M. (1976a). Similarities in the polypeptide composition of glyoxysomal and endoplasmic reticulum membranes from castor bean endosperm. *Biochem. J.* **154,** 491–499.

Bowden, L., and Lord, J. M. (1976b). The cellular origin of glyoxysomal proteins in germinating castor bean endosperm. *Biochem. J.* **154,** 501–506.

Bowden, L., and Lord, J. M. (1978). Purification and comparative properties of microsomal and glyoxysomal malate synthetase from castor bean endosperm. *Plant Physiol.* **61,** 237–245.

Breidenbach, R. W., and Beevers, H. (1967). Association of glyoxylate cycle enzymes in a novel subcellular particle from castor bean endosperm. *Biochem. Biophys. Res. Commun.* **27,** 462–469.

Briggs, R. T., Draht, D. B., Karnovsky, M. L., and Karnowsky, M. J. (1975). Localization of NADH oxidase of the surface of human polymorphonuclear leukocytes by a new cytochemical method. *J. Cell Biol.* **67,** 566–586.

Bringer, S., Bochem, H., Sprey, B., and Sahm, H. (1980). Microbodies in the brown rot fungus *Poria contigua*. *Zentralbl. Bakteriol. Hyg.* **1,** 193–199.

Brody, M., and White, J. E. (1972). Environmental factors controlling enzymatic activity in microbodies and mitochondria of *Euglena gracilis*. *FEBS Lett.* **23,** 149–152.

Brown, R. H., and Merrett, M. J. (1977). Density labelling during microbody development in cotyledons. *New Phytol.* **79,** 73–81.

Brown, R. H., Lord, J. M., and Merrett, M. J. (1974). Fractionation of the proteins of plant microbodies. *Biochem. J.* **144,** 559–566.

Bullock, K. W., Deason, T. R., and O'Kelley, J. C. (1979). Occurrence of glycolate oxidase in some coccoid, zoospore-producing green algae. *J. Phycol.* **15,** 142–146.

Burke, J. J., and Trelease, R. N. (1975). Cytochemical demonstration of malate synthase and glycolate oxidase in microbodies in cucumber cotyledons. *Plant Physiol.* **56,** 710–717.

Chabot, J. F., and Chabot, B. F. (1974). Microbodies in conifer needle mesophyll. *Protoplasma* **79,** 349–358.

Chang, C. C., and Beevers, H. (1968). Biogenesis of oxalate in plant tissues. *Plant Physiol.* **43,** 1821–1828.

Chang, C. C., and Huang, A. H. C. (1981). Metabolism of glycolate in isolated spinach leaf peroxisomes. Kinetics of glyoxylate, oxalate, CO_2 and glycine formation. *Plant Physiol.* **67,** 1003–1006.

Cheesbrough, T. M., and Moore, T. S. (1980). Transverse distribution of phospholipids in organelle membranes from *Ricinus communis* L. var. Hale endosperm. *Plant Physiol.* **65,** 1076–1080.

Ching, T. M. (1970). Glyoxysomes in megagametophyte of germinating Ponderosa pine seeds. *Plant Physiol.* **46,** 475–583.

Ching, T. M. (1973). Compartmental utilization of carboxyl-^{14}C-tripalmitin by tissue homogenates of pine seeds. *Plant Physiol.* **51,** 278–284.

Choinski, J. S., and Trelease, R. N. (1978). Control of enzyme activities in cotton cotyledons during maturation and germination. *Plant Physiol.* **62,** 141–145.

Chua, N. H., and Schmidt, G. W. (1978). Post translational transport into intact chloroplasts of a precursor to the small subunit of ribulose-1, 5-diphosphate carboxylase. *Proc. Natl. Acad. Sci. U.S.A.* **75,** 6110–6114.

Cioni, M., Pinzauti, G., and Vanni, P. (1981). Comparative biochemistry of the glyoxylate cycle. *Comp. Biochem. Physiol. B* **70B,** 1–26.

Codd, G. A., and Schmid, G. H. (1972). Enzymic evidence for peroxisomes in a mutant of *Chlorella vulgaris. Arch. Mikrobiol.* **81,** 264–272.

Colby, J., Dalton, H., and Whittenbury, R. (1979). Biological and biochemical aspects of microbial growth on C_1 compounds. *Annu. Rev. Microbiol.* **33,** 481–517.

Collins, N., and Merrett, M. J. (1975a). The localization of glycollate-pathway enzymes in *Euglena. Biochem. J.* **148,** 321–328.

Collins, N., and Merrett, M. J. (1975b). Microbody marker-enzymes during transition from phototrophic to organotrophic growth in *Euglena. Plant Physiol.* **55,** 1018–1022.

Cooper, R. A., and Lloyd, D. (1972). Subcellular fractionation of the colourless alga *Polytomella caeca* by differential and zonal centrifugation. *J. Gen. Microbiol.* **72,** 59–70.

Cooper, T. G. (1971). The activation of fatty acids in castor bean endosperm. *J. Biol. Chem.* **246,** 3451–3455.

Cooper, T. G., and Beevers, H. (1969a). Mitochondria and glyoxysomes from castor bean endosperm. Enzyme constituents and catalytic capacity. *J. Biol. Chem.* **244,** 3507–3513.

Cooper, T. G., and Beevers, H. (1969b). β-oxidation in glyoxysomes from castor bean endosperm. *J. Biol. Chem.* **244,** 3514–3520.

Coulthard, C. E., Michaelis, R., Short, W. F., Sykes, G., Skrimshire, G. E. H., Standfast, A. F. B., Birkinshaw, J. H., and Raistrick, H. (1945). Notatin: an antibacterial glucose-aerodehydrogenase from *Penicillium notatum* Westling and *Penicillium resticulosum* sp. nov. *Biochem J.* **39,** 24–36.

Cronshaw, J. (1964). Crystal containing bodies of plant cells. *Protoplasma* **59,** 318–325.

Curti, L. B., Ronchi, S., Branzoli, U., Ferri, G., and Williams, C. H. (1973). Improved purification, amino acid analysis and molecular weight of homogeneous D-amino acid oxidase from pig kidney. *Biochim. Biophys. Acta* **327,** 266–273.

Davis, B., and Merrett, M. J. (1975). The glycolate pathway and photosynthetic competence in *Euglena. Plant Physiol.* **55,** 30–34.

DeBoer, J., and Feierabend, J. (1974). Comparison of the effects of cytokinins on enzyme development in different cell compartments of the shoot organs of rye seedlings. *Z. Pflanzenphysiol.* **71,** 261–270.

de Duve, C. (1965). Functions of microbodies (peroxisomes). *J. Cell Biol.* **27,** 25A.

de Duve, C., and Baudhuin, P. (1966). Peroxisomes (microbodies and related particles). *Physiol. Rev.* **46,** 323–357.

Deisseroth, A., and Dounce, A. L. (1970). Catalase: Physical and chemical properties, mechanism of catalysis, and physiological role. *Physiol. Rev.* **50,** 319–375.

DeMaggio, A. E., Greene, C., and Stetler, D. (1980). Biochemistry of fern spore germination. Glyoxylate and glycolate cycle activity in *Onoclea sensibilis* L. *Plant Physiol.* **66,** 922–924.

Doman, D. C., Walker, J. C., Trelease, R. N., and Moore, B. (1982). Metabolism of carbohydrate and lipid reserves in germinating cotton seeds. *Planta* (in press).

Donaldson, R. P., and Beevers, H. (1977). Lipid composition of organelles from germinating castor bean endosperm. *Plant Physiol.* **59,** 259–263.

Donaldson, R. P., Tolbert, N. E., and Schnarrenberger, C. (1972). A comparison of microbody membranes with microsomes and mitochondria from plant and animal tissue. *Arch. Biochem. Biophys.* **152,** 199–215.

Douglass, S. A., Criddle, R. S., and Breidenbach, R. W. (1973). Characterization of deoxyribonucleic acid species from castor bean endosperm. Inability to detect a unique deoxyribonucleic acid species associated with glyoxysomes. *Plant Physiol.* **51,** 902–906.

Edwards, G. E., and Huber, S. C. (1981). The C_4 pathway. *In* "The Biochemistry of Plants" (P. K. Stumpf and E. E. Conn, eds.), Vol. 8, pp. 238–281. Academic Press, New York.

Fahimi, H. D., and Yokota, S. (1981). Ultrastructural and cytochemical aspects of animal peroxisomes - some recent observation. *In* "International Cell Biology, 1980–1981" (H. G. Schweiger, ed.), pp. 640–650. Springer-Verlag, Berlin and New York.

Feierabend, J. (1975). Developmental studies on microbodies in wheat leaves. III. On the photocontrol of microbody development. *Planta* **123,** 63–77.

Feierabend, J., and Beevers, H. (1972a). Developmental studies on microbodies in wheat leaves. I. Conditions influencing enzyme development. *Plant Physiol.* **49,** 28–32.

Feierabend, J., and Beevers, H. (1972b). Developmental studies on microbodies in wheat leaves. II. Ontogeny of particulate enzyme associations. *Plant Physiol.* **49,** 33–39.

Feierabend, J., and Brassel, D. (1977). Subcellular localization of shikimate dehydrogenase in higher plants. *Z. Pflanzenphysiol.* **82,** 334–346.

Feierabend, J., and Schrader-Reichardt, V. (1976). Biochemical differentiation of plastids and other organelles in rye leaves with a high-temperature-induced deficiency of plastid ribosomes. *Planta* **129,** 133–145.

Floyd, G. L., and Salisbury, J. L. (1977). Glycolate dehydrogenase in primitive green algae. *Am. J. Bot.* **64,** 1294–1296.

Frederick, S. E., and Newcomb, E. H. (1969a). Cytochemical localization of catalase in leaf microbodies (peroxisomes). *J. Cell Biol.* **43,** 343–353.

Frederick, S. E., and Newcomb, E. H. (1969b). Microbody-like organelles in leaf cells. *Science* **163,** 1353–1355.

Frederick, S. E., and Newcomb, E. H. (1971). Ultrastructure and distribution of microbodies in leaves of grasses with and without CO_2-photorespiration. *Planta* **96,** 152–174.

Frederick, S. E., Newcomb, E. H., Vigil, E. L., and Wergin, W. P. (1968). Finestructural characterization of plant microbodies. *Planta* **8,** 229–252.

Frederick, S. E., Gruber, P. J., and Tolbert, N. E. (1973). The occurrence of glycolate dehydrogenase and glycolate oxidase in green plants: An evolutionary survey. *Plant Physiol.* **52,** 318–323.

Frederick, S. E., Gruber, P. J., and Newcomb, E. H. (1975). Plant microbodies. *Protoplasma* **84,** 1–29.

Frevert, J., and Kindl, H. (1980a). Purification of glyoxysomal acetyl-CoA acyltransferase. *Hoppe-Seyler's Z. Physiol. Chem.* **361,** 537–542.

Frevert, J., and Kindl, H. (1980b). A bifunctional enzyme from glyoxysome-purification of a protein possessing enoyl-CoA hydratase and 3-hydroxyacyl-CoA dehydrogenase. *Eur. J. Biochem.* **107,** 79–86.

Frevert, J., Köller, W., and Kindl, H. (1980). Occurrence and biosynthesis of glyoxysomal enzymes in ripening cucumber seeds. *Hoppe-Seyler's Z. Physiol. Chem.* **361,** 1557–1565.

Frey-Wyssling, A., Grieshaber, E., and Mühlethaler, K. (1963). Origin of spherosomes in plant cells. *J. Ultrastruct. Res.* **8,** 506–516.

Fridovich, I. (1975). Superoxide dismutase. *Annu. Rev. Biochem.* **44,** 147–159.

Frigerio, N. A., and Harbury, H. A. (1958). Preparation and some properties of crystalline glycolic acid oxidase. *J. Biol. Chem.* **231,** 135–157.

Fukui, S., and Tanaka, A. (1979). Yeast peroxisomes. *Trends Biochem. Sci. (Pers. Ed.)* **4,** 246–249.

Gahan, P. B. (1968). Lysosomes. *In* "Plant Cell Organelles" (J. B. Pridham, ed.), pp. 228–238. Academic Press, New York.

Gemmrich, A. R. (1981). Ultrastructural and enzymatic studies on the development of microbodies in germinating spores of the fern. *Anemia phyllitidis. Z. Pflanzenphysiol.* **102,** 69–80.

Gerhardt, B. (1971). Localization of microbodial enzymes in *Polytomella caeca, Arch. Mikrobiol.* **80,** 205–218.

Gerhardt, B. (1973). Untersuchungen zur Funktionsänderung der Microbodies in den Keimblättern von *Helianthus annuus. Planta* **110,** 15–28.

Gerhardt, B. (1978). Microbodies/Peroxisomen pflanzlicher Zellen. "Cell Biology Monographs," Vol. 5, 283 pp. Wien:Springer.

Gerhardt, B. (1981). Enzyme activities of the β-oxidation pathway in spinach leaf peroxisomes. *FEBS Lett.* **126,** 71–73.

Gerhardt, B., and Beevers, H. (1969). Occurrence of RNA in glyoxysome from castor bean endosperm. *Plant Physiol.* **44,** 1475–1477.

Gibbs, D. A., and Watts, R. W. E. (1966). An investigation of the possible role of xanthine oxidase in the oxidation of glyoxylate to oxalate. *Clin. Sci.* **31,** 285–297.

Gibbs, M. (1969). Photorespiration, Warburg effect and glycolate. *Ann. N. Y. Acad. Sci.* **168,** 356–368.

Godovari, H. R., Badour, S. S., and Waygood, E. R. (1973). Isocitrate lyase in green leaves. *Plant Physiol.* **51,** 863–867.

Goldman, B. M., and Blobel, G. (1978). Biogenesis of peroxisomes: Intracellular site of synthesis of catalase and uricase. *Proc. Natl. Acad. Sci. U.S.A.* **75,** 5066–5070.

Gonzalez, E. (1978). Effect of gibberellin A_3 on the endoplasmic reticulum and on the formation of glyoxysomes in the endosperm of germinating castor bean. *Plant Physiol.* **62,** 449–453.

Gonzalez, E. (1982). Aggregated forms of malate and citrate synthase are localized in endoplasmic reticulum of endosperm of germinating castor bean. *Plant Physiol.* **69,** 83–87.

Gonzalez, E., and Beevers, H. (1976). Role of the endoplasmic reticulum in glyoxysome formation in castor bean endosperm. *Plant Physiol.* **57,** 406–409.

Goodman, D. B. P., Davis, W. L., and Jones, R. G. (1980). Glyoxylate cycle in toad urinary bladder: Possible stimulation by aldosterone. *Proc. Natl. Acad. Sci. U.S.A.* **77,** 1521–1525.

Graves, L. B., Jr., and Becker, W. M. (1974). Beta-oxidation in glyoxysomes from *Euglena. J. Protozool.* **21,** 771–773.

Graves, L. B., Jr., Trelease, R. N., Grill, A., and Becker, W. M. (1972). Localization of glyoxylate cycle enzymes in glyoxysomes in *Euglena. J. Protozool.* **19,** 527–532.

Graves, L. B., Jr., Armentrout, V. N., and Maxwell, D. P. (1976). Distribution of glyoxylate-cycle enzymes between microbodies and mitochondria in *Asperigillus tamarii. Planta* **132,** 143–148.

Gregor, H. D. (1977). A new method for the rapid separation of cell organelles. *Anal. Biochem.* **82,** 255–258.

Gregory, R. P. (1968). An improved preparative method for spinach catalase and evaluation of some of its properties. *Biochim. Biophys. Acta* **159,** 429–439.

Grodzinski, B., and Butt, V. S. (1977). The effect of temperature on glycollate decarboxylation in leaf peroxisomes. *Planta* **133,** 261–266.

Gruber, P. J., and Frederick, S. E. (1977). Cytochemical localization of glycolate oxidase in microbodies of *Klebsormidium*. *Planta* **135**, 45–49.

Gruber, P. J., Becker, W. M., and Newcomb, E. H. (1973). The development of microbodies and peroxisomal enzymes in greening bean leaves. *J. Cell Biol.* **56**, 500–518.

Hanks, J. F., Tolbert, N. E., and Schubert, K. R. (1981). Localization of enzymes of ureide biosynthesis in peroxisomes and microsomes of nodules. *Plant Physiol.* **68**, 65–69.

Hanzely, L., and Vigil, E. L. (1975). Fine structural and cytochemical analysis of phragmosomes (microbodies) during cytokinesis in *Allium* root tip cells. *Protoplasma* **86**, 269–277.

Hayashi, H., Taya, K., Suga, T., and Ninobe, S. (1976). Studies on peroxisomes. VI. Relationship between peroxisomal core and urate oxidase. *J. Biochem.* **79**, 1029–1034.

Hebant, C., and Marty, F. (1972). Fine structural identification of peroxisomes in the cells of the photosynthetic lamellae of the leaf of *Polytrichum commune* gametophytes. *J. Bryol.* **7**, 195–199.

Herzog, V., and Fahimi, H. D. (1974). The effect of glutaraldehyde on catalase. Biochemical and cytochemical studies with beef liver catalase and rat liver peroxisomes. *J. Cell Biol.* **60**, 303–311.

Hicks, D. B., and Donaldson, R. P. (1982). Electron transport in glyoxysomal membranes. *Arch. Biochem. Biophys.* **215**, 280–288.

Hock, B., and Gietl, C. (1982). Intracellular transport of malate dehydrogenase isoenzymes. *Ann. N. Y. Acad. Sci.* **386**, 350–360.

Hong, Y. N., and Schopfer, P. (1981). Control by phytochrome of urate oxidase and allantoinase activities during peroxisome development in the cotyledons of mustard (*Sinapis alba* L.) seedlings. *Planta* **152**, 325–335.

Horner, H. Y., Jr., and Arnott, H. J. (1966). A histochemical and ultrastructural study of pre- and post-germinated yucca seeds. *Bot. Gaz. (Chicago)* **127**, 48–64.

Huang, A. H. C. (1975). Comparative studies of glyoxysomes from various fatty seedlings. *Plant Physiol.* **55**, 870–874.

Huang, A. H. C. (1983). Plant lipases. *In* "Lipolytic Enzymes" (B. Borgstrom and H. Brockman, eds.), Elsevier/North-Holland Biomedical Press, Amsterdam (in press).

Huang, A. H. C., and Beevers, H. (1971). Isolation of microbodies from plant tissues. *Plant Physiol.* **48**, 637–641.

Huang, A. H. C., and Beevers, H. (1972). Microbody enzymes and carboxylases in sequential extracts from C_4 and C_3 leaves. *Plant Physiol.* **50**, 242–248.

Huang, A. H. C., and Beevers, H. (1973). Localization of enzymes within microbodies. *J. Cell Biol.* **58**, 379–389.

Huang, A. H. C., and Beevers, H. (1974). Developmental changes in endosperm of germinating castor bean independent of embryonic axis. *Plant Physiol.* **54**, 277–279.

Huang, A. H. C., Bowmann, P. D., and Beevers, H. (1974). Immunological and biochemical studies on isozymes of malate dehydrogenase and citrate synthetase in castor bean glyoxysomes. *Plant Physiol.* **54**, 364–368.

Huang, A. H. C., Liu, K. D. F., and Youle, R. J. (1976). Organelle-specific isozymes of aspartate-alpha-ketoglutarate transaminiase in spinach leaves. *Plant Physiol.* **58**, 110–113.

Hutton, D., and Stumpf, P. K. (1969). Characterization of the β-oxidation systems from maturing and germinating castor bean seeds. *Plant Physiol.* **44**, 508–516.

Hutton, D., and Stumpf, P. K. (1971). The pathway of ricinoleic acid catabolism in the germinating castor bean and pea. *Arch. Biochem. Biophys.* **142**, 48–60.

Inestrosa, N. C., Bronfman, M., and Leighton, F. (1980). Purification of the peroxisomal fatty acyl-CoA oxidase from rat liver. *Biochem. Biophys. Res. Commun.* **95**, 7–12.

Itaya, K., Fukumoto, J., and Yamamoto, T. (1971). Studies on urate oxidase of *Candida utilis*. II. Some physical and chemical properties of the purified enzyme. *Agric. Biol. Chem.* **35,** 813–821.

Jacks, T. J., Yatsu, L. Y., and Altschul, A. M. (1967). Isolation and characterization of peanut spherosomes. *Plant Physiol.* **42,** 585–597.

Jensen, R. G., and Bahr, J. T. (1977). Ribulose 1,5-biphosphate carboxylase-oxygenase. *Annu. Rev. Plant Physiol.* **28,** 379–400.

Jones, R. G., Davis, W. L., and Goodman, D. B. P. (1981). Microperoxisomes in the epithelial cells of the amphibian urinary bladder: An electron microscopic demonstration of catalase and malate synthase. *J. Histochem. Cytochem.* **29,** 1150–1156.

Kagawa, T., and Beevers, H. (1975). The development of microbodies (glyoxysomes and leaf peroxisomes) in cotyledons of germinating seedlings. *Plant Physiol.* **55,** 258–264.

Kagawa, T., and Gonzalez, E. (1981). Organelle specific isozymes of citrate synthase in the endosperm of developing *Ricinus* seedlings. *Plant Physiol.* **68,** 845–850.

Kagawa, T., Lord, J. M., and Beevers, H. (1973a). The origin and turnover of organelle membranes in castor bean endosperm. *Plant Physiol.* **51,** 61–65.

Kagawa, T., McGregor, K., and Beevers, H. (1973b). Development of enzymes in the cotyledons of watermelon seedlings. *Plant Physiol.* **51,** 66–71.

Kagawa, T., Lord, J. M., and Beevers, H. (1975). Lecithin synthesis during microbody biogenesis in watermelon cotyledons. *Arch. Biochem. Biophys.* **167,** 45–53.

Kashiwagi, K., Tobe, T., and Higashi, T. (1971). Studies on rat liver catalase. V. Incorporation of ^{14}C-leucine into catalase by isolated rat liver ribosomes. *J. Biochem.* **70,** 785–793.

Kates, M., and Marshall, M. O. (1975). Biosynthesis of phosphoglycerides in plants. *Rec. Adv. Chem. Biochem. Plant Lipids, Proc. Symp., 1974* pp. 115–159.

Kawamoto, S., Tanaka, A., Yamamura, M., Teranishi, Y., and Fukui, S. (1977). Microbody of n-alkane-grown yeast. Enzyme localization in the isolated microbody. *Arch. Microbiol.* **112,** 1–8.

Kawamoto, S., Yamada, T., Tanaka, A., and Fukui, S. (1979). Distinct subcellular localization of NAD-linked and FAD-linked glycerol-3-phosphate dehydrogenases in n-alkane-grown *Candida tropicalis. FEBS Lett.* **97,** 253–256.

Kerr, W. M., and Groves, D. (1975). Purification and properties of glycollate oxidase from *Pisum sativum* leaves. *Phytochemistry* **14,** 359–362.

Khan, F. R., Saleemuddin, M., Siddiqi, M., and McFadden, B. H. (1977). Purification and properties of isocitrate lyase from flax seedlings. *Arch. Biochem. Biophys.* **183,** 13–23.

Kindl, H. (1982). Glyoxysome biogenesis via cytosolic pools. *Ann. N. Y. Acad. Sci.* **386,** 314–326.

Kindl, H., and Ruis, H. (1971). Metabolism of aromatic amino acids in glyoxysomes. *Phytochemistry* **10,** 2633–2636.

Kindl, H., Köller, W., and Frevert, J. (1980). Cytosolic precursor pools during glyoxysome biogenesis. *Z. Physiol. Chem.* **361,** 465–467.

Kisaki, T., and Tolbert, N. E. (1969). Glycolate and glyoxylate metabolism by isolated peroxisomes or chloroplasts. *Plant Physiol.* **44,** 242–250.

Kluge, M., and Ting, I. P. (1978). "Crassulacean Acid Metabolism." Springer-Verlag, Berlin and New York.

Kobr, M. J., Vanderhaeghe, F., and Combepine, G. (1973). Particulate enzymes of the glyoxylate cycle in *Neurospora crassa. Biochem. Biophys. Res. Commun.* **37,** 460–645.

Kohn, L. D., and Warren, W. A. (1970). The kinetic properties of spinach leaf glyoxylic acid reductase. *J. Biol. Chem.* **245,** 3831–3839.

Köller, W., and Kindl, H. (1977). Glyoxylate cycle enzymes of the glyoxysomal membrane from cucumber cotyledons. *Arch. Biochem. Biophys.* **181,** 236–248.

Köller, W., and Kindl, H. (1978). Studies supporting the concept of glyoxyperoxisomes as intermediary organelles in transformation of glyoxysomes into peroxisomes. *Z. Naturforsch. C: Biosci.* **33C,** 962–968.

Köller, W., and Kindl, H. (1980). 19S cytosolic malate synthase. A small pool characterized by rapid turnover. *Hoppe-Seyler's. Z. Physiol. Chem.* **361,** 1437–1444.

Köller, W., Frevert, J., and Kindl, H. (1979). Incomplete glyoxysomes appearing at a late stage of maturation of cucumber seeds. *Z. Naturforsch. C: Biosci.* **34C,** 1232–1236.

Kornberg, H. L., and Beevers, H. (1957). The glyoxylate cycle as a stage in the conversion of fat to carbohydrate in castor beans. *Biochim. Biophys. Acta.* **26,** 531–537.

Kudielka, R., Kock, H., and Theimer, R. R. (1981). Substrate dependent formation of glyoxysomes in cell suspension cultures of anise (*Pimpinella anisum* L.). *FEBS Lett.* **136,** 8–12.

Lamb, J. E., Riezman, H., Leaver, C. J., and Becker, W. M. (1978). Regulation of glyoxysomal enzymes during germination of cucumber. *Plant Physiol.* **62,** 754–760.

Lazarow, P. B. (1981). Functions and biogenesis of peroxisomes. 1980. *In* "International Cell Biology, 1980–1981" (H. G. Schweiger, ed.), pp. 633–639. Springer-Verlag, Berlin and New York.

Lazarow, P. B., Mariette, R., Fujiki, Y., and Wong, L. (1982). Biosynthesis of peroxisomal proteins *in vivo* and in *in vitro. Ann. N. Y. Acad. Sci.* **386,** 285–298.

LeBault, J. M., Roche, B., Duvnjak, Z., and Azoulay, E. (1970). Alcool et aldehydes deshydrogenases particularies de Candida tropicalis cultive sur hydrocarbures. *Biochim. Biophys. Acta* **220,** 373–385.

Liang, Z., Yu, C., and Huang, A. H. C. (1982). Isolation of spinach leaf peroxisomes in 0.25 M sucrose solution by Percoll density gradient centrifugation. *Plant Physiol.* **70,** 1210–1212.

Lin, Y. H., Moreau, R. A., and Huang, A. H. C. (1982). Involvement of glyoxysomal lipase in the hydrolysis of storage triacylglycerol in the cotyledons of soybean seedlings. *Plant Physiol.* **70,** 108–112.

Lips, S. H. (1975). Enzyme content of plant microbodies as affected by experimental procedures. *Plant Physiol.* **55,** 598–601.

Liu, K. D. F., and Huang, A. H. C. (1976). Developmental studies of NAD-malate dehydrogenase isozymes in the cotyledon of cucumber seedlings grown in darkness and in light. *Planta* **131,** 279–284.

Liu, K. D. F., and Huang, A. H. C. (1977). Subcellular localization and development changes of aspartate-α-ketoglutarate transaminase isozymes in the cotyledons of cucumber seedlings. *Plant Physiol.* **59,** 777–782.

Lode, E. T., and Coon, M. J. (1973). Role of rubredoxin in fatty acid and hydrocarbon hydroxylation reactions. *In* "Iron-Sulfur Proteins" (W. Lovenberg, ed.), pp. 173–191. Academic Press, New York.

Longo, G. P., and Longo, C. P. (1975). Development of mitochondrial enzyme activities in germinating maize scutellum. *Plant Sci. Lett.* **5,** 339–346.

Longo, C. P., Bernasconi, E., and Longo, P. G. (1975). Solubilization of enzymes from glyoxysomes of maize scutellum. *Plant Physiol.* **55,** 1115–1119.

Longo, G. P., Pedretti, M., Rossi, G., and Longo, C. P. (1979). Effect of benzyladenine on the development of plastids and microbodies in excised watermelon cotyledons. *Planta* **145,** 209–217.

Lopes-Perez, M. J., Gimenes-Solves, A., Calonge, F. D., and Santos-Ruiz, A. (1974). Evidence of glyoxysomes in germinating pine seeds. *Plant Sci. Lett.* **2,** 377–386.

Lord, J. M. (1978). Evidence that a proliferation of the endoplasmic reticulum precedes the formation of glyoxysomes and mitochondria in germinating castor bean endosperm. *J. Exp. Bot.* **29,** 13–23.

Lord, J. M. (1980). Biogenesis of peroxisomes and glyoxysomes. *Subcell. Biochem.* **7,** 171–211.

Lord, J. M., and Beevers, H. (1972). The problem of reduced nicotinamide adenine dinucleotide oxidation in glyoxysomes. *Plant Physiol.* **49,** 249–251.

Lord, J. M., and Roberts, L. M. (1982). Glyoxysome biogenesis via the endoplasmic reticulum? *Ann. N. Y. Acad. Sci.* **386,** 362–374.

Lowens, F. (1980). L-Ascorbic acid: Metabolism, biosynthesis, function. *In* "The Biochemistry of Plants" (P. K. Stumpf and E. E. Conn, eds.), Vol. 3, pp. 77–99. Academic Press, New York.

Lück, H. (1965). Catalase. *In* "Methods of Enzymatic Analysis" (H. U. Bergmeyer, ed.), PP. 885–894. Academic Press, New York.

Ludwig, B., and Kindl, H. (1976). Plant microbody proteins. II. Purification and characterization of the major protein component (SP-63) of peroxisome membranes. *Hoppe-Seyler's Physiol. Chem.* **357,** 177–186.

Maccecchini, M. L., Rudin, Y., Blobel, G., and Schatz, G. (1979). Import of proteins into mitochondria: Precursor forms of the extra mitochondrially made F_1-ATPase subunits in yeast. *Proc. Natl. Acad. Sci. U.S.A.* **76,** 343–347.

McCord, J. M. (1979). Superoxide dismutase: Occurrence, structure, function, and evolution. *Isozymes: Curr. Top. Biol. Med. Res.* **3,** 1–21.

McKinley, M. P., and Trelease, R. N. (1980). Regulation of carbon flow through the glyoxylate and tricarboxylic acid cycles in the mitochondria of *Turbatrix aceti.* I. Coarse and fine controls. *Comp. Biochem. Physiol. B* **67B,** 17–26.

Manton, I. (1961). Observations on phragmosomes. *J. Exp. Bot.* **12,** 108–113.

Marriott, K. M., and Northcote, D. H. (1977). The influence of abscisic acid, adenosine 3′, 5′ cyclic phosphate, and gibberellic acid on the induction of isocitrate lyase activity in the endosperm of germinating castor bean seeds. *J. Exp. Bot.* **28,** 219–224.

Massey, V., Palmer, G., and Bennett, R. (1961). The purification and some properties of D-amino acid oxidase. *Biochim. Biophys. Acta* **48,** 1–9.

Masters, C., and Holmes, R. (1977). Peroxisomes: New aspects of cell physiology and biochemistry. *Physiol. Rev.* **57,** 816–877.

Matile, P. H. (1975). "The Lytic Compartment of Plant Cells," p. 183. Springer-Verlag, Berlin and New York.

Maxwell, D. P., and Bateman, D. F. (1968). Oxalic acid biosynthesis by *Sclerotium rolfsii. Phytopathology* **58,** 1635–1642.

Maxwell, D. P., Maxwell, M. D., Hänssler, G., Armentrout, V. N., Murray, G. M., and Hoch, H. C. (1975). Microbodies and glyoxylate-cycle enzyme activities in filamentous fungi. *Planta* **124,** 109–123.

Maxwell, D. P., Armentrout, V. N., and Graves, L. B., Jr. (1977). Microbodies in plant pathogenic fungi. *Annu. Rev. Phytopathol.* **15,** 119–134.

Mellor, R. B., Bowden, L., and Lord, J. M. (1978). Glycoproteins of the glyoxysomal matrix. *FEBS Lett.* **90,** 275–278.

Mettler, J. J., and Beevers, H. (1980). Oxidation of NADH in glyoxysomes by a malate aspartate shuttle. *Plant Physiol.* **66,** 555–560.

Miernyk, J. A., and Trelease, R. N. (1981a). Control of enzyme activities in cotton cotyledons during maturation and germination. IV. β-Oxidation. *Plant Physiol.* **67,** 341–346.

Miernyk, J. A., and Trelease, R. N. (1981b). Role of malate synthase in citric acid synthesis by maturing cotton embryos - A proposal. *Plant Physiol.* **67,** 875–881.

Miernyk, J. A., and Trelease, R. N. (1981c). Substrate specificity of cotton glyoxysomal enoyl-CoA hydratase. *FEBS Lett.* **129,** 139–141.

Miernyk, J. A., and Trelease, R. N. (1981d). Malate synthase from *Gossypium hirsutum*. *Phytochemistry* **20,** 2657–2663.

Miernyk, J. A., Trelease, R. N., and Choinski, J. S., Jr. (1979). Malate synthase activity in cotton and other ungerminated oilseeds. A survey. *Plant Physiol.* **63,** 1068–1071.

Miflin, B., and Beevers, H. (1974). Isolation of intact plastids from a range of plant tissues. *Plant Physiol.* **53,** 870–874.

Mills, G. L., and Cantino, E. C. (1975). The single microbody in the zoospore of *Blastocladiella emersonii* is a "Symphyomicrobody". *Cell Differ.* **4,** 35–44.

Mishina, M., Kamiryo, T., Tashiro, S., and Numa, S. (1978). Separation and characterization of two long-chain acyl CoA synthetase from *Candida lipolytica*. *Eur. J. Biochem.* **82,** 347–354.

Mollenhauer, H. H., Morré, J. D., and Kelly, A. G. (1966). The widespread occurrence of plant cytosomes resembling animal microbodies. *Protoplasma* **62,** 44–52.

Moore, T. S. (1976). Phosphatidycholine synthesis in castor bean endosperm. *Plant Physiol.* **57,** 383–386.

Moore, T. S. (1982). Phospholipid biosynthesis. *Annu. Rev. Plant Physiol.* **33,** 235–359.

Moreau, R. A., and Huang, A. H. C. (1977). Gluconeogenesis from storage wax in the cotyledons of *Jojoba* seedlings. *Plant Physiol.* **60,** 329–333.

Moreau, R. A., and Huang, A. H. C. (1979). Oxidation of fatty alcohols in the cotyledons of *jojoba* seedlings. *Arch. Biochem. Biophys.* **194,** 422–430.

Moreau, R. A., Liu, K. D. F., and Huang, A. H. C. (1980). Spherosomes in castor bean endosperm. Membrane components, formation, and degradation. *Plant Physiol.* **65,** 1176–1180.

Müller, M. (1975). Biochemistry of protozoan microbodies. *Annu. Rev. Microbiol.* **29,** 467–483.

Muto, S., and Beevers, H. (1974). Lipase activities in castor bean endosperm. *Plant Physiol.* **54,** 23–28.

Neat, C. E., Thomassen, M. S., and Osmundsen, H. (1980). Induction of peroxisomal β-oxidation in rat liver by high-fat diets. *Biochem. J.* **186,** 369–371.

Nelson, E. B., and Tolbert, N. E. (1970). Glycolate dehydrogenase in green algae. *Arch. Biochem. Biophys.* **141,** 102–110.

Neuburger, M., Journet, E. P., Bligny, R., Cadre, J. P., and Douce, R. (1982). Purification of plant mitochondria by isopycnic centrifugation in density gradients of Percoll. *Arch. Biochem. Biophys.* **217,** 312–323.

Newcomb, E. H., and Frederick, S. E. (1971). Distribution and structure of plant microbodies (peroxisomes). *In* "Photosynthesis and Photorespiration" (M. D. Hatch, C. B. Osmond, and R. O. Slatyer, eds.), pp. 452–457. Wiley (Interscience), New York.

Newcomb, E. H., and Tandon, S. R. (1981). Uninfected cells of soybean root nodules: Ultrastructure suggests key role in ureide production. *Science* **212,** 1394–1396.

Nishimura, M., and Beevers, H. (1978). Hydrolases in vacuoles from castor bean endosperm. *Plant Physiol.* **62,** 44–48.

Nishimura, M., Graham, D., and Akazawa, T. (1976). Isolation of intact chloroplasts and other cell organelles from spinach leaf protoplasts. *Plant Physiol.* **58,** 309–314.

Novikoff, A. B., and Goldfischer, S. (1969). Visualization of peroxisomes (microbodies) and mitochondria with diaminobenzidine. *J. Histochem. Cytochem.* **17,** 675–680.

Novikoff, P. M., and Novikoff, A. B. (1972). Peroxisomes in absorptive cells of mammalian small intestine. *J. Cell Biol.* **53,** 532–560.

Oaks, A., and Beevers, H. (1964). The glyoxylate cycle in maize scutellum. *Plant Physiol.* **39,** 431–434.

Oliver, D. J. (1981). Role of glycine and glyoxylate decarboxylation in photorespiratory CO_2 release. *Plant Physiol.* **68,** 1031–1934.
Ory, R. L., Yatsu, L. Y., and Kircher, H. W. (1968). Association of lipase activity with the spherosomes of *Ricinus communis. Arch. Biochem. Biophys.* **264,** 255–264.
Osumi, M., and Kazama, H. (1978). Microbody-associated DNA in *Candida tropicalis* pK 233 cells. *FEBS Lett.* **90,** 309–312.
Pais, M. S., and Carrapico, F. (1979). Localisation cytochimique de la malate synthétase et de la glycolate oxydase au niveau des microbodies des spores chlorophylliennes de la mousse Bryum capillare. *C. R. Acad. Sci. Ser. D.* **288,** 395–398.
Palade, G. E. (1975). Intracellular aspects of the process of protein secretion. *Science* **189,** 347–358.
Parish, R. (1971). The isolation of peroxisomes, mitochondria and chloroplasts from leaves of spinach beet (*Beta vulgaris* L. ssp. vulgaris). *Eur. J. Biochem.* **22,** 423–429.
Patel, T. R., and McFadden, B. A. (1977). Particulate isocitrate lyase and malate synthase in *Caenorhabditis elegans. Arch. Biochem. Biophys.* **183,** 24–30.
Paul, J. S., Sullivan, C. W., and Volcani, B. E. (1975). Photorespiration in diatoms. Mitochondrial glycolate dehydrogenase in *Cylindrotheca fusiformis* and *Nitzschia alba. Arch. Biochem. Biophys.* **169,** 152–159.
Penner, D., and Ashton, F. M. (1967). Hormonal control of isocitrate lyase synthesis. *Biochem. Biophys. Acta* **148,** 481–485.
Porter, K. R., and Caulfield, J. B. (1958). The formation of the cell plate during cytokinesis in *Allium cepa* L. *Proc. Int. Conf. Electron Microsc., IV* **2,** 503–509.
Powell, M. J. (1976). Ultrastructure and isolation of glyoxysomes (microbodies) in zoospores of the fungus *Entophyetis* sp. *Protoplasma* **89,** 1–27.
Powell, M. J. (1978). Phylogenetic implications of the microbody-lipid globule complex in zoosporic fungi. *BioSystems* **10,** 167–180.
Rawsthorne, S., Minchin, F. R., Summerfield, R. J., Cookson, C., and Coombs, J. (1980). Carbon and nitrogen metabolism in legume root nodules. *Phytochemistry* **19,** 341–355.
Redman, C. B., Grab, D. J., and Irukulla, R. (1972). The intracellular pathway of newly formed rat liver catalase. *Arch. Biochem. Biophys.* **152,** 496–501.
Rehfeld, D. W., and Tolbert, N. E. (1972). Aminotransferases in peroxisomes from spinach leaves. *J. Biol. Chem.* **247,** 4803–4811.
Rhodin, J. (1954). Correlation of ultrastructural organization and function in normal and experimentally changed proximal convoluted tubule cells of the mouse kidney. *Aktiebolaget Godvil, Stockholm.*
Richardson, K. E., and Tolbert, N. E. (1961). Oxidation of glyoxylic acid to oxalic acid by glycolic acid oxidase. *J. Biol. Chem.* **231,** 1280–1284.
Rocha, V., and Ting, I. P. (1970). Tissue distribution of microbody, mitochondria and soluble MDH isoenzymes. *Plant Physiol.* **46,** 754–756.
Roggenkamp, R., Sahm, H., Hinkelmann, W., and Wagner, F. (1975). Alcohol oxidase and catalase in peroxisomes of methanol-grown *Candida boidinii. Eur. J. Biochem.* **59,** 231–236.
Rothe, G. (1974). Intracellular compartmentation and regulation of two shikimate dehydrogenase isoenzymes in *Pisum sativum. Z. Pflanzenphysiol.* **74,** 152–159.
Rouiller, C., and Bernhard, W. (1956). "Microbodies" and the problem of mitochondrial regeneration in liver cells. *J. Biophys. Biochem. Cytol.* **2** (suppl), 355–359.
Sahm, H., and Wagner, F. (1973). Microbial assimilation of methanol. The ethanol- and methanol-oxidizing enzymes of the yeast *Candida boidinii. Eur. J. Biochem.* **36,** 250–256.

Sawaki, S., Hattori, N., Morikawa, N., and Yamada, K. (1967). Oxidation and reduction of glyoxylate by lactate dehydrogenase. *J. Vitaminol.* **13,** 93–97.

Scandalios, J. G. (1975). Differential gene expression and biochemical properties of catalase isozymes in maize. *In* "Isozymes. III. Developmental Biology" (C. L. Markert, ed.), pp. 213–238. Academic Press, New York.

Scandalios, J. G., Tong, W. F., and Roupakias, D. G. (1980a). Cat 3, a third gene locus coding for a tissue-specific catalase in maize: Genetics, intracellular location, and some biochemical properties. *Mol. Gen. Genet.* **179,** 33–41.

Scandalios, J. G., Chang, D. Y., McMillin, D. E., Tsaftaris, A., and Moll, R. H. (1980b). Genetic regulation of the catalase developmental program in maize scutellum: Identification of a temporal regulatory gene. *Proc. Natl. Acad. Sci. U.S.A.* **77,** 5360–5364.

Schmitt, M. R., and Edwards, G. E. (1982). Isolation and purification of intact peroxisomes from green leaf tissue. *Plant Physiol.* **70,** 1213–1217.

Schnarrenberger, C., and Burkhard, C. (1977). *In-vitro* interaction between chloroplasts and peroxisomes as controlled by inorganic phosphate. *Planta* **134,** 109–114.

Schnarrenberger, C., Oeser, A., and Tolbert, N. E. (1971). Development of microbodies in sunflower cotyledons and castor bean endosperm during germination. *Plant Physiol.* **48,** 566–574.

Schopfer, P., Bajracharya, D., Falk, H., and Thien, W. (1975). Phytochromgesteuerte Entwicklung von Zellorganellen (Plastiden, Microbodies, Mitochondrien). *Ber. Dtsch. Bot. Ges.* **88,** 245–268.

Schopfer, P., Bajracharya, D., Bergfield, R., and Falk, H. (1976a). Phytochrome-mediated transformation of glyoxysomes into peroxisomes in the cotyledons of mustard (*Sinapis alba* L.) seedlings. *Planta* **133,** 73–80.

Schopfer, P., Bajracharya, D., and Falk, H. (1976b). Photocontrol of microbody and mitochondrion development: The involvement of phytochrome. *In* "Light and Plant Development" (H. Smith, ed.), pp. 193–212. Butterworth, Boston.

Servettaz, O., Cortesi, F., and Longo, C. P. (1976). Effect of benzyladenine on some enzymes of mitochondria and microbodies in excised sunflower cotyledons. *Plant Physiol.* **58,** 569–572.

Shnitka, T. K., and Talibi, G. G. (1971). Ctyochemical localization of ferricyanide reduction of α-hydroxacid oxidase activity in peroxisomes of rat kidney. *Histochemistry* **27,** 137–158.

Silverberg, B. A. (1975). An ultrastructural and cytochemical characterization of microbodies in the green algae. *Protoplasma* **83,** 269–295.

Sjogren, R. E., and Romano, A. H. (1967). Evidence for multiple forms of isocitrate lyase in *Neurospora crassa. J. Bacteriol.* **93,** 1638–1643.

Somerville, C. R., and Ogren, W. L. (1982). Genetic modification of photorespiration. *Trends Biol. Sci.* **7,** 171–174.

Sorenson, J. C., and Scandalios, J. G. (1980). Biochemical characterization of a catalase inhibitor from maize. *Plant Physiol.* **66,** 688–691.

Stabenau, H. (1974). Verteilung von microbody-enzymen aus *Chlamydomonas* in Dichtegradienten. *Planta* **118,** 35–42.

Stabenau, H. (1976). Microbodies from *Spirogyra.* Organelles of a filamentous alga similar to leaf peroxisomes. *Plant Physiol.* **58,** 693–695.

Stabenau, H., and Beevers, H. (1974). Isolation and characterization of microbodies from the alga *Chlorogonium elongatum. Plant Physiol.* **53,** 866–869.

Stabenau, H., and Säftel, W. (1982). A peroxisomal glycolate oxidase in the algae *Mougeotia. Planta* **154,** 165–167.

Stewart, C. R., and Beevers, H. (1967). Gluconeogenesis from amino acids in germinating castor bean endosperm and its role in transport to the embryo. *Plant Physiol.* **42,** 1587–1595.

Stewart, K. D., and Mattox, K. R. (1978). Structural evolution in the flagellated cells of green algae and land plants. *Biosystems* **10,** 145–152.

Sullivan, J. O., and Casselton, P. J. (1972). The subcellular localization of glyoxylate cycle enzymes in *Corpinus lagopus* (sensu Buller). *J. Gen. Microbiol.* **75,** 333–337.

Surendranathan, K. K., and Nair, P. M. (1978). Purification and characterization of a natural inhibitor for isocitrate lyase, present in gamma-irradiated preclimacteric banana. *Plant Sci. Lett.* **12,** 169–175.

Susani, M., Zimniak, P., Fessl, F., and Ruis, H. (1976). Localization of catalase A in vacuoles of *Saccharomyces cerevisia:* Evidence for the vacuolar nature of isolated "yeast peroxisomes". *Hoppe-Seyler's Z. Physiol. Chem.* **357,** 961–970.

Szabo, A., and Avers, C. J. (1969). Some aspects of regulation of peroxisomes and mitochondria in yeast. *Ann. N. Y. Acad. Sci.* **168,** 302–312.

Tajima, S., and Yamamoto, Y. (1975). Enzymes of purine catabolism in soybean plants. *Plant Cell Physiol.* **16,** 271–282.

Tanaka, A., Yasuhara, S., Kawamoto, S., Fukui, S., and Osumi, M. (1976). Development of microbodies in the yeast *Kloeckera* growing on methanol. *J. Bacteriol.* **126,** 919–927.

Tanaka, A., Osumi, M., and Fukui, S. (1982). Peroxisomes of alkane-grown yeast: Fundamental and practical aspects. *Ann. N. Y. Acad. Sci.* **386,** 183–198.

Theimer, R. R. (1976). A specific inactivator of glyoxysomal isocitrate lyase from sunflower cotyledons. *FEBS Lett.* **62,** 297–300.

Theimer, R. R., and Beevers, H. (1971). Uricase and allantoinase in glyoxysomes. *Plant Physiol.* **47,** 246–251.

Theimer, R. R., and Heidinger, P. (1974). Control of particulate urate oxidase activity in bean roots by external nitrogen supply. *Z. Pflanzenphysiol.* **73,** 360–370.

Theimer, R. R., and Rosnitschek, I. (1978). Development and intracellular localization of lipase activity in rapeseed (*Brassica napus* L.) cotyledons. *Planta* **139,** 249–256.

Theimer, R. R., and Theimer, E. (1975). Studies on the development and localization of catalase and H_2O_2 generating oxidases in the endosperm of germinating castor bean. *Plant Physiol.* **56,** 100–104.

Theimer, R. R., Anding, G., and Schmid-Neuhaus, B. (1975). Density labeling evidence against a de novo formation of peroxisomes during greening of fat-storing cotyledons. *FEBS Lett.* **57,** 89–92.

Theimer, R. R., Anding, G., and Matzner, P. (1976). Kinetin action on the development of microbody enzymes in sunflower cotyledons in the dark. *Planta* **128,** 41–47.

Theimer, R. R., Wanner, G., and Anding, G. (1978). Isolation and biochemical properties of two types of microbody from *Neurospora crassa* cells. *Cytobiologie* **18,** 132–144.

Thomas, R. J., and Schrader, L. E. (1981). Ureide metabolism in higher plants. *Phytochemistry* **20,** 361–371.

Thomas, J., and Trelease, R. N. (1981). Cytochemical localization of glycolate oxidase in microbodies (glyoxysomes and peroxisomes) of higher plant tissues with the $CeCl_3$ technique. *Protoplasma* **108,** 39–53.

Thomas, S. M., ap Rees, T. (1972). Gluconeogenesis during the germination of *Cucurbita pepo. Phytochemistry* **11,** 2177–2185.

Thornton, R. M., and Thimann, K. V. (1964). On a crystal-containing body in cells of the oat coleoptile. *J. Cell Biol.* **20,** 345–350.

Ting, I. P., Fuhr, I., Curry, R., and Zschoche, W. C. (1975). Malate dehydrogenase isozymes in plants: Preparation, properties, and biological significance. *In* "Isozymes.

II. Physiological Function (C. L. Markert, ed.), pp. 369–386. Academic Press, New York.

Tolbert, N. E. (1971). Microbodies-peroxisomes and glyoxysomes. *Annu. Rev. Plant Physiol.* **22,** 45–74.

Tolbert, N. E. (1980). Microbodies-peroxisomes and glyoxysomes. *In* "The Biochemistry of Plants" (P. K. Stumpf and E. E. Conn, eds.), pp. 359–388. Academic Press, New York.

Tolbert, N. E. (1981). Metabolic pathways in peroxisomes and glyoxysomes. *Annu. Rev. Biochem.* **50,** 133–157.

Tolbert, N. E., Oeser, A., Kisaki, T., Hageman, R. H., and Yamazaki, R. K. (1968). Peroxisomes from spinach leaves containing enzyme related to glycolate metabolism. *J. Biol. Chem.* **243,** 5179–5184.

Tolbert, N. E., and Essner, E. (1981). Microbodies: Peroxisomes and glyoxysomes. *J. Cell Biol.* **91,** 271s–283s.

Trelease, R. N. (1975). Malate synthase. *In* "Electron Microscopy of Enzymes" (M. A. Hyat, ed.), Vol. 4, pp. 158–176. Van Nostrand-Reinhold, New York.

Trelease, R. N., Becker, W. M., Gruber, P. J., and Newcomb, E. H. (1971). Microbodies (glyoxysomes and peroxisomes) in cucumber cotyledons. Correlative biochemical and ultrastructural study in light- and dark-grown seedlings. *Plant Physiol.* **48,** 461–475.

Trelease, R. N., Becker, W. M., and Burke, J. J. (1974). Cytochemical localization of malate synthase in glyoxysomes. *J. Cell Biol.* **60,** 483–495.

Tsaftaris, A., and Scandalios, J. G. (1981). Regulation of glyoxysomal enzyme expression in maize. *Differentiation (Berlin).* **18,** 133–140.

Tsubouchi, J., Tonomoura, K., and Tanaka, K. (1976). Ultrastructure of microbodies of methanol-assimilating yeasts. *J. Gen. Appl. Microbiol.* **22,** 131–142.

van Dijken, J. P., and Bos, P. (1981). Utilization of amines by yeast. *Arch. Microbiol.* **128,** 320–324.

van Dijken, J. P., and Veenhuis, M. (1980). Cytochemical localization of glucose oxidase in peroxisomes of *Asperigillus niger. Eur. J. Appl. Microbiol. Biotechnol.* **9,** 275–283.

van Dijken, J. P., Veenhuis, M., and Harder, W. (1982). Peroxisomes of methanol-grown yeasts. *Ann. N. Y. Acad. Sci.* **386,** 200–215.

Veenhuis, M., van Dijken, J. P., and Harder, W. (1976). Cytochemical studies on the localization of methanol oxidase and other oxidases in peroxisomes of methanol-grown *Hansenula polymorpha. Arch. Microbiol.* **111,** 123–135.

Veenhuis, M., van Dijken, J. P., Pilon, S. A., and Harder, W. (1978). Development of crystalline peroxisomes in methanol-grown cells of the yeast *Hansenula polymorpha* and its relation to environmental conditions. *Arch. Microbiol.* **117,** 153–163.

Veenhuis, M., Keizer, I., and Harder, W. (1979). Characterization of peroxisomes in glucose-grown *Hansenula polymorpha* and their development after transfer of cells into methanol-containing media. *Arch. Microbiol.* **120,** 167–175.

Veenhuis, M., van Dijken, J. P., and Harder, W. (1982). The significance of peroxisomes in the metabolism of one-carbon compounds in yeasts. *Adv. Microbial. Physiol.* (in press).

Vigil, E. L. (1970). Cytochemical and developmental changes in microbodies (glyoxysomes) and related organelles of castor bean endosperm. *J. Cell Biol.* **46,** 435–454.

Vigil, E. L. (1973). Structure and function of plant microbodies. *Subcell. Biochem.* **2,** 237–285.

Vogels, G. D., and van der Drift, C. (1976). Degradation of purines and pyrimidines by microorganisms. *Bacteriol. Rev.* **40,** 403–468.

Walk, R. A., and Hock, B. (1977). Glyoxysomal malate dehydrogenase of watermelon cotyledons: *De novo* synthesis on cytoplasmic ribosomes. *Planta* **134,** 277–285.

Walk, R. A., and Hock, B. (1978). Cell-free synthesis of glyoxysomal malate dehydrogenase. *Biochem. Biophys. Res. Commun.* **81,** 636–643.

Walker, J. D., and Cooney, J. J. (1973). Pathway of n-alkane oxidation in *Cladosporium resinae. J. Bacteriol.* **115,** 635–639.

Walton, N. J. (1982). Glyoxylate decarboxylation during glycollate oxidation by pea leaf extracts: significance of glyoxylate and extract concentrations. *Planta* **155,** 218–224.

Wanner, G., and Theimer, R. R. (1978). Membranous appendices of spherosomes (oleosomes). Possible role in fat utilization in germinating oilseeds. *Planta* **140,** 163–169.

Wanner, G., and Theimer, R. R. (1982). Two types of microbodies in *Neurospora crassa. Ann. N. Y. Acad. Sci.* **386,** 269–282.

White, J. E., and Brody, M. (1974). Enzymatic characterization of sucrose-gradient microbodies of dark-grown, greening and continuously light-grown *Euglena gracilis. FEBS Lett.* **40,** 325–330.

Wickner, W. (1979). Assembly of proteins into membranes. *Science* **210,** 861–868.

Woodward, J., and Merrett, M. J. (1975). Induction potential for glyoxylate cycle enzymes during the cell cycle of *Euglena gracilis. Eur. J. Biochem.* **55,** 555–559.

Wrigley, A., and Lord, J. M. (1977). The effects of gibberellic acid on organelle biogenesis in the endosperm of germinating castor bean seeds. *J. Exp. Bot.* **28,** 345–353.

Yamada, H., Adachi, O., and Ogata, K. (1965). Amine oxidases of microorganisms. Part III. Properties of amine oxidase of *Aspergillus niger. Agric. Biol. Chem.* **29,** 864–869.

Yamada, M., Tanaka, T., Kader, J. C., and Mazliak, P. (1978). Transfer of phospholipids from microsomes to mitochondria in germinating castor bean endosperm. *Plant Cell Physiol.* **19,** 173–176.

Yamada, T., Nawa, H., Kawamoto, S., Tanaka, A., and Fukui, S. (1980). Subcellular localization of long-chain alcohol dehydrogenase and aldehyde-dehydrogenase in n-alkane-grown *Candida tropicalis. Arch. Microbiol.* **128,** 145–151.

Yokota, A., Nakano, Y., and Kitaoka, S. (1978a). Different effects of some growing condition on glycolate dehydrogenase in mitochondria and microbodies in *Euglena gracilis. Agric. Biol. Chem.* **42,** 115–120.

Yokota, A., Nakano, Y., and Kitaoka, S. (1978b). Metabolism of glycolate in mitochondria of *Euglena gracilis. Agric. Biol. Chem.* **42,** 121–129.

Yoshimura, T., Isemura, T. (1971). Subunit structure of glucose oxidase from *Penicillium amagasakiense. J. Biochem.* **69,** 839–846.

Youle, R. J., and Huang, A. H. C. (1976). Development and properties of fructose 1,6-bisphosphatase in the endosperm of castor bean seedlings. *Biochem. J.* **154,** 647–652.

Young, O., and Beevers, H. (1976). Mixed function oxidases from germinating castor bean endosperm. *Phytochemistry* **15,** 378–385.

Zelitch, I. (1971). "Photosynthesis, Photorespiration, and Plant Productivity." Academic Press, New York.

Zelitch, I. (1972). The photoxidation of glyoxylate by envelop-free spinach chloroplasts and its relation to photorespiration. *Arch. Biochem. Biophys.* **150,** 698–707.

Zimmermann, R., and Neupert, W. (1980). Biogenesis of glyoxysomes. Synthesis and intracellular transfer of isoitrate lyase. *Eur. J. Biochem.* **112,** 225–233.

Zindler-Frank, E. (1976). Oxalate biosynthesis in relation to photosynthetic pathway and plant productivity - a survey. *Z. Pflanzenphysiol.* **80,** 1–13.

Zwart, K., Veenhuis, M., Dijken, J. P., and Harder, W. (1980). Development of amine oxidase-containing peroxisomes in yeast during growth on glucose in the presence of methylamine as the sole source of nitrogen. *Arch. Microbiol.* **126,** 117–126.

Index

H